Rosalie Bertell has a doctorate in biometrics and has worked in the field of environmental health since 1969. She has been involved in the founding of several organisations, including the International Institute of Concern for Public Health in Toronto, Canada, of which she is President. She led the Bhopal and Chernobyl Medical Commissions, has undertaken collaborative research with numerous organisations and is the recipient of the Right Livelihood Award, the World Federalist Peace Prize, the United Nations Environment Programme (UNEP) Global 500 Award and five honorary doctorates. Dr Bertell is a member of a Roman Catholic religious congregation, the Grey Nuns of the Sacred Heart.

Also by Dr Rosalie Bertell from The Women's Press:

No Immediate Danger (1985)

PLANET EARTH
THE LATEST WEAPON OF WAR

ROSALIE BERTELL

BLACK ROSE BOOKS

Montréal/New York/London

Black Rose Books No. EE296

Hardcover ISBN: 1-55164-183-6 (bound)
Paperback ISBN: 1-55164-182-8 (pbk.)

Canadian Cataloguing in Publication Data

Bertell, Rosalie, 1929-
Planet earth : the newest weapon of war

Includes bibliographical references and index.
Hardcover ISBN: 1-55164-183-6 (bound)
Paperback ISBN: 1-55164-182-8 (pbk.)

1. War--Environmental aspects. 2. Military weapons--Environmental aspects.
3. Space weapons--Environmental aspects. I. Title.

UG1520.B47 2001   303.6'6   C00-901400-4

Cover design by Associés libres, Montréal

BLACK ROSE BOOKS

C.P. 1258	2250 Military Road	99 Wallis Road
Succ. Place du Parc	Tonawanda, NY	London, E9 5LN
Montréal, H2W 2R3	14150	England
Canada	USA	UK

To order books in North America:
(phone) 1-800-565-9523 (fax) 1-800-221-9985

In Europe: (phone) London 44 (0)20 8986-4854 (fax) 44 (0)20 8533-5821

Our Web Site address: http://www.web.net/blackrosebooks

A publication of the Institute of Policy Alternatives of Montréal (IPAM)

Printed in Canada

The Canada Council | Le Conseil des Arts
for the Arts | du Canada

*To the many wonderful people who have mentored
and awakened me*

ACKNOWLEDGEMENTS

I would like to gratefully acknowledge the companionship of my Sisters, the Grey Nuns of the Sacred Heart, who gifted me with the time and space to develop this analysis. I am indebted also for the financial support of the Right Livelihood Foundation, Nancy's Very Own Foundation, and the many supporters of our Institute. The staff and Board of the International Institute of Concern for Public Health have provided me with time, moral support and the comfort of knowing that this institutional effort at providing new approaches to holistic community health will continue.

There are many persons, forming a new layer of civil society, who are dedicated to the transformation of institutions and the formation of a truly collegial global solidarity. They will find many of their passions, hopes and insights in harmony with my analysis. I thank them for their lives, their work for peace and their generous sharing of constructive creativity. It will be this network of global thinkers and planners, rather than the bankers and corporate executives, who offer the best non-violent future for our planet.

However, it is to Earth itself that I owe the most gratitude. It has strongly attracted me and engaged my respect and wonder all of my life.

CONTENTS

INTRODUCTION

I felt the piercing cold and saw the clear blue sky and magnificent sun. It was an unreal experience, this winter day at the top of Beckley Hill in Vermont. I was more used to a winter filled with overcast, cold and dreary days, and had formed a close mental association between sunny and warm.

The sunny cold of Vermont made me think more generally about deceptive appearances and how misleading a 'first look' can be. My mother always looked well, even at the age of 95, and this was probably because of the twinkle in her eyes and the fact that her spirit was still fully alive. Some of my friends with cancer could have walked in a beauty pageant and no one would ever have noticed that they were sick. It made me think about the Earth and the delicately balanced natural processes that regulate it. If the Earth were damaged or suffering from some 'illness', would we be able to recognise the problem early on, when it might be possible to reverse the process?

On this Vermont day, the birch trees were stripped of their leaves, standing naked in seasonal repose. But this bareness was normal, natural, and in the spring, the delicate green leaves would appear again to clothe the trees in elegance. Obviously one must understand the complete life cycle of a natural organism in order not to mistake a dormant period for death or deterioration. The Earth itself has cycles, and our human ancestors have faithfully marked the passing of the seasons and the weather for some 150 years. However, our knowledge of how these cycles function and how they interact is, as yet, incomplete. We do not know how resilient the Earth is, nor can we gauge its capacity to heal itself.

On a clear cold day, the Earth looks wonderful, the air feels refreshing and it can be hard to believe the warnings that we have seriously compromised its health. Yet, since the United Nations

Conference on the Environment in 1972, it has become obvious that the Earth faces serious problems: trees dying, species becoming extinct, contamination and depletion of drinking water, soil erosion, deforestation, smog, reduction of fish stocks, poverty and overcrowding. More recently, the incidence of violent weather has been increasing at an alarming rate and there is evidence that many so-called 'natural' disasters are linked to human activities. All of our attempts to restore the health of the planet by changing our lifestyle, reducing dependence on fossil fuels, 'reusing, recycling and reducing' seem not to have stemmed the tide. In fact in September 1999 the United Nations Environment Program announced that the environmental crisis is deepening not receding.

It is my belief that we have been treating the symptoms but not the cause of the disease of the Earth. We have been abusing Earth's natural systems, the way it regulates temperature and water supply, recycles waste and protects life. For me, some of the most fundamental abuses have occurred because of our continued reliance on the military.

Wars result in immediate deaths and destruction, but the environmental consequences can last hundreds, often thousands of years. And it is not just war itself that undermines our life support system, but also the research and development, military exercises and general preparation for battle that are carried out on a daily basis in most parts of the world. The majority of this pre-war activity takes place without the benefit of civilian scrutiny and therefore we are unaware of some of what is being done to our environment in the name of 'security'.

While there is a legitimate need for a police force in the global community, there can be no rationale for a military force. Blowing up a neighbourhood suspected of harbouring a criminal has never been seen as a civilised way of promoting domestic order. Nor is destruction of a nation and contamination of its food, air and other resources a means of achieving global peace. Of course, the inability to wage war does not eliminate regional disputes – it merely guarantees that the disputes will be submitted to negotiated settlement rather than violence. Large political and trade coalitions, such as the Organization of African States (OAS) and the European Union, can be formed through legal discussions rather than coming together through force.

In fact, I believe that our definition of global security has become outdated. Military security has its foundations in either the protection of wealth, land and privilege or the desire to confiscate the wealth and land of others. Modern society seems to have an unhealthy dependency on economic gain, and this has resulted in a widening gap between the haves and the have nots of the world. This is a major destabilising factor that actually causes global insecurity, not security. It also distorts a market economy towards catering for the wealthy whilst the needs of the poor go unanswered. This cannot be the basis of true democracy.

I also believe we have been confused by the struggle between communism and capitalism, which has been the dominant dialogue among thinkers for many years. This is basically a conflict over how to manage the excess in an economy. The essence of the dispute is whether accumulation of wealth should be held by government, which claims to use it for the benefit of the masses through funding of social programmes, or by private entrepreneurs, who think they can more wisely 'build the economy' thereby providing jobs and a better standard of living for the people.

The problem with both systems is that they have focused on economic stability at the expense of ecological and social stability, when it has become increasingly clear that these three are inter-dependent. The most urgent problem facing us at the moment is how to sustain Earth, our life-support system, not how to redistribute wealth (although I think if we learn how to do the former we will be forced to recognise the wisdom of the latter). A meeting of the G7 to decide on interest rates cannot rectify the over-stretching of our natural resources and the manipulation of Earth's restorative power! Life thrives on balance, not on a singular focus on the economic 'bottom line'.

However, this goal of balanced social planning requires that we first provide a new job description for the military in order that they truly fulfil their purpose of serving and protecting the interests of the people. In order to do this, we have to look beyond the model of global dominance through force towards more gentle, cooperative solutions to the problems we face. To many, this may seem idealistic in a world dominated by what I would call a hard and unbending capitalism. But it is only by envisaging ideal solutions that we can begin the process of change.

Already there are signs of hope. The women's movement and the growing awareness of human rights, animal rights and Earth rights are all signs of profound transformations in societal structures and in the way inequalities and conflict are addressed. The United Nations is undergoing a period of reform and can now benefit from fifty years of experience. The stewardship of the land exercised for centuries by indigenous people is slowly being recognised and their ability to live in the midst of plenty without exploiting or destroying that abundance is a lesson for those who aspire to global management. The present crisis is a global one and in order to solve it, we must seek global solutions.

This book is divided into three parts. In part one, I will examine two major conflicts that give a snapshot of high-tech war at the end of the twentieth century. In the portrait of these two wars, I hope to give the reader some idea of the extensive environmental impact of modern weapons and also to question the motivation and results of what has been called 'humanitarian' intervention. I also hope to demonstrate that in understanding our history, we can better see the implications of today's preparations for war and find a new path into the future.

But war itself is only one side of the military coin. Equally destructive to the health of our planet is the military experiment-ation and research which exploits our natural resources and destabilises a balanced ecology. In part two I will look at some of the consequences of past research and show how the tendency to experiment first and ask questions later characterises the search for ever-more sophisticated weaponry, especially in the race towards 'Star Wars'. It is worth noting that in this book I focus primarily on the exploitation of the Earth environment as a weapon, although my analysis of the overall problem does of course include atomic, biological and chemical weapons. This is because there is more public awareness of the dangers of these 'a, b, c' weapons, a popular consensus that they are unacceptable, and international legislation prohibiting their use. The fact that many countries continue to develop them only serves to emphasise the need to find new means of solving international disputes.

In the final section of *Planet Earth*, I seek to redefine our notion of security. At present, the greatest threat to our security is not invasion by 'the enemy'; it is the destruction of the natural resources upon which we all rely for life and health. Without efficient use and responsible management of these resources, the fabric of civilisation will disintegrate and we will be reduced to fighting with each other over basics such as clean air and water. In order to provide future generations with what I have called 'ecological' security, we need to work on both a global and a local level.

The reader will not find every detail of current military strategy, military exercises and hardware in this book. Equally, the environmental problems may not be explained in the depth required for an environmental scientist. Social policy experts will probably find the selection of hopeful signs and directions lacking in detail and omitting things they consider to be important. But my purpose here is not to concentrate on one particular speciality. This book is a broad-brush picture of the totality and my goal is to set a direction of discourse and scholarship which is different from that of mainstream linear thinking. It is an appeal for interdisciplinary strategies, because I see the single focus of the military and the environmental or social scientist as one of the greatest menaces to the survival of the planet. If the environmentalists begin to look at the impact and implications of military exercises, the international policy experts begin to think about the survival of the planet, and the military strategists realise that they are capable of making this beautiful Earth unlivable, then my book will have met its goal.

But it is not just the 'experts' who have a role to play – we all need to rethink our responses to the main problems of the day in order to avoid ecological collapse and we must adopt practices that will be beneficial to the future of our planet.

One of the greatest barriers to civilian action is the secrecy that surrounds military projects. In working through this book, I sometimes felt overwhelmed with anger at the abuse of our Earth and sadness for our widespread ignorance of the affairs and undertakings of our own countries and allies. It should not be necessary to have to explain in detail all of the intrigue and ill-conceived experiments which have been undertaken by the governments of democratic countries. Responsibility for national actions and policy rests with the people, and keeping the people ignorant undermines the very meaning of democracy.

However, it is not only the civilian community who have been kept in the dark. Young politicians today know almost nothing about the atmospheric nuclear testing in the 1950s, and less than nothing about the history of military ionospheric experiments. (I'm sure very few politicians choose the *Astrophysics Journal* as bedtime reading!) Therefore we do not have an historical context in which to understand the present; and we do not have the language to interpret military plans for the future.

It is my hope that this book will open up for the reader an historical matrix against which to view the present and future and a dictionary of terms with which to understand current military strategy. I also hope that it will spur the reader to become involved in peaceful enterprises. We must set up a cooperative relationship with the Earth, not one of dominance, for it is ultimately the gift of life that we pass on to our children and the generations to follow.

PART I
WARFARE

CHAPTER 1
WAR IN THE LAST TEN YEARS OF THE 20TH CENTURY

Those of us who have lived through the years since the dropping of the first nuclear bomb thought that the end of the Cold War spelled a return to sanity. For over fifty years, an uneasy peace had been maintained through the fear of nuclear retaliation and the deadly competition between East and West. Following the fall of the Berlin Wall, the collapse of the Soviet Union, and the introduction of *perestroika*, the threat of war seemed to recede and the global community breathed a sigh of relief. But instead of demobilisation, the last decade of the twentieth century saw two markedly one-sided Western wars, one against Iraq, the other against the former Yugoslavia.

It is imperative that we analyse these post-Cold War events in order to decide whether or not they were preventable and to determine what impact they have had on global society and planet Earth. There are invaluable lessons we can learn about the causes of modern war: how it can be avoided and what the real after-effects are. Looking back we can see how the roots of a conflict often became obscured. Looking forward, we can find clues in past wars as to the nature and possible results of future conflict. Weaponry is becoming increasingly lethal to humans and disruptive to our life-support system, and if we fail to prevent such violent outbreaks, our prospects will look bleak indeed.

The purpose, manner of conduct, rules of engagement and language of war have changed radically over the past ten years. There have, of course, been wars in Rwanda, Sierra Leone and Indonesia (to name but a few), but both the Kosovan crisis and the Gulf War were remarkable in that they were presented to the public as 'humanitarian' causes; they were 'corrective wars', punishing two nations for unacceptable behaviour. The West set itself up as judge and jury; it decided upon the sentence and was responsible for its execution.

Currently in matters of war and peace, there are two major players: the North Atlantic Treaty Organization (NATO) and the United Nations (UN) Security Council. The war against Iraq was undertaken with the consent of the UN Security Council; the air strikes in Kosovo were initiated by NATO, an organisation supposedly committed only to defence of its member countries, without United Nations approval. Indeed in June 2000, a UK foreign affairs select committee concluded that NATO had no powers under its treaty to conduct a 'humanitarian operations war' without the specific authority of the United Nations.[1] Clearly there was confusion over who could decide on questions of global security.

Democratic decision-making requires an objective and unbiased viewpoint, but serious questions can be raised over the neutrality of these two organisations. Within the UN Security Council decision-making power is unequally balanced. The five nuclear nations – the US, the UK, France, China and Russia – are permanent members and have the power of veto, whilst the other ten members are elected every two years by the General Assembly. NATO, a voluntary coalition of 19 nations, has even less neutrality and democratic power-sharing. Dominance in decision-making depends heavily on the financial shares taken on by the member nations, with the US normally assuming the lion's share. It also depends on the cutting-edge weaponry being used, and again the US dominates in new weapon designs and fire power. Formerly strong international players, like the UK, Germany and France, now play a supporting role in policies developed in Washington.

This inequality begins to make sense if we look at US military policy. In a major policy document, called 'The Defense Planning Guide' (1992), the Pentagon describes US foreign policy in the post-Cold War period.[2] It blatantly stated that the only way forward must be to strengthen and expand US political and military dominance, adding that no other country had the right to aspire to a role of leadership, even as a regional power.

Our first objective is to prevent the re-emergence of a new rival. First, the US must show the leadership necessary to establish and protect a new world order that holds the promise of convincing potential competitors that they need not aspire to a greater role or pursue a more aggressive posture to protect their legitimate interests.

We must account sufficiently for the interests of advanced industrial nations to discourage them from challenging our leadership or seeking to overturn the established political and economic order. Finally, we must maintain the mechanism for deterring potential competitors from even aspiring to a larger regional or global role.[3]

This is the rule of the gun, not the rule of law. It brings to mind the Wild West, when the 'strongest' male assumed leadership and meted out justice as he saw fit until someone 'bigger' de-throned him. There is a clear equation between the domination of the United States and established political and economic order. Bearing this in mind, it is not unreasonable to suggest that the imbalance in the decision-making powers of the UN Security Council and NATO helps protect the US's lofty status.

This is the climate being set for international decision making at the beginning of the new millennium, an arena in which accountability and justice for all nations is being decided upon by a select few. Of course, when a nation behaves inhumanely or oppresses its people, the international community cannot be expected just to stand by. As the conflicts in Kosovo and the Gulf demonstrate, however, notions of who is 'guilty' and who is 'right' are not always so clear. Both wars were humanitarian and environmental disasters. In such a context, nobody wins.

THE KOSOVAN CRISIS

The roots of the Kosovan crisis can be traced far back into history but the economic crisis that precipitated the break-up of Yugoslavia is closely linked to the intervention of the International Monetary Fund (IMF) in the 1980s. Yugoslavia required 'credits' to allow it to purchase goods on the international market, and in 1986 the IMF began to tie these credits to political reform and constitutional change. Over time, the IMF effectively took over economic policy in Yugoslavia, under the advice and influence of economists from

Harvard University and the Massachusetts Institute of Technology (MIT). The IMF insisted that in return for further credits, the Yugoslav government had to open its economy to full foreign ownership rights and put an end to worker participation, moving the country towards a Western-style economy.

In the aftermath of the Cold War, so called 'shock therapy' for rapidly moving a country from socialism to capitalism was in vogue. The Yugoslav government was pressured to 'come into the twentieth century' – to raise taxes, negotiate international loans, scale down social programmes and stop funding transfer payments to the Yugoslavian republics. This last move in particular caused tensions in a country that, until this point, had been held together by its socialist arrangement, including subsidies for education, health care, social services, and the transferring of money from richer provinces to areas most in need.

The effect of this shock therapy on the economy was devastating. The Yugoslavian dinar, which was worth $22 in 1986, dropped to $0.11 by December 1989. By December 1991, hyperinflation had set in. Increasingly onerous conditions further dislocated the economy. As the largest contributor to the IMF, the US insisted on a brutal structural-adjustment austerity programme: devalue the currency, freeze wages, cut all subsidies, close many state-run industries and privatise others, and increase unemployment to 20 per cent.[4]

Then, in November 1990, the US Congress passed what became the 1991 Foreign Operations Appropriations Law, which suddenly, and without warning, cut off all US aid and credits. This law demanded that separate democratic elections take place within the next six months in each of the six republics that made up Yugoslavia. The US policy made it clear that aid would only be resumed after these elections, which had to be approved by the State Department. This dramatic move was presented to the public as a 'humanitarian' issue, but it is clear that it had as much to do with the conversion of Yugoslavia into a capitalist country with a market economy in the wake of the Cold War. As *NATO Review* stated in May 1996, 'The European Community and NATO are involved in anchoring the countries of Eastern Europe ... in order to consolidate the gains of the Cold War.'

Without credit, Yugoslavia could not purchase raw materials or trade internationally to gain currency with which to pay its debts. The US action sent a clear message to Europe that if the Balkan republics could not repay their debts, they would be forced to declare bankruptcy, and Yugoslavia's real assets would be up for grabs. The European Community followed the US lead, suspending its economic aid, imposing an embargo on Yugoslavian weapons imports and insisting that it hold multi-party elections or face economic blockade.

On 5 May 1991, the six-month deadline imposed by the US passed. It was apparent that the socialist economic system that had held together the fragile Yugoslavian federation was in tatters. When the IMF shock therapy hit, Croatia and Slovenia, the richer republics, had had to bear a greater burden to help the poorer republics. Massive and repeated strikes and other labour action proved ineffective against the international economic forces of change. Political and economic tensions seemed to spark old nationalistic antagonisms between Serbs, Croats and Muslims. The country collapsed in ethnic fighting, genocide and chaos. On 25 June 1991, Slovenia and Croatia declared independence, and civil war began. In April the unrest erupted in Bosnia, and the suffering and bloodshed there was on a scale Europe had not seen since World War II.

This is necessarily a simplified version of an immensely complex series of events, but what is salient is that in the midst of political turmoil, the IMF withheld credits which Yugoslavia needed in order to be able to trade. A European economist told me that the resulting chaos 'had to happen'; the change from socialism to capitalism was inevitable. He justified his views by saying that the Europeans had told Yugoslavia not to set up their socialist system in the first place. 'We have won the cold war, things have to change!'

But Europe could hardly stand by and watch a country in its own backyard plunge headlong into civil war. So in March 1992, Yugoslavia and the European Community brokered a deal in Lisbon, Portugal. The three main parties – Serbs, Croats and Muslims – agreed to Swiss-style cantonisation of Yugoslavia, with self-determination in each of the three cantons. The plan also called for European supervision of Yugoslavia during its transition to a market economy. However, after an intervention by US Ambassador Zimmermann, the Muslim leader, President Ljubo Izetbegovic, reversed his support and backed out of the deal. He was soon followed by the Croatian representative, Mate Boban, with the result that the plan was never implemented.[5]

In May 1993, US Secretary of State Cyrus Vance and former British Foreign Secretary Lord David Owen, the United Nations and European Community representatives respectively, signed the Vance Owen plan, which recommended dividing Yugoslavia into ten provinces. Again, Europe was to take the lead in assisting Yugoslavia to solve its economic and political problems. Owen stated publicly that Washington undermined the deal after it was negotiated. The actions of the US are only understandable if we refer to the words of the previously quoted 1992 Pentagon 'Defense Planning Guide': 'We must seek to prevent the emergence of European-only security arrangements which would undermine NATO.'

The Dayton Accord (1995), supposedly the centrepiece of Balkan diplomacy, was in essence the same as the two earlier peace agreements, differing only in that it was to be implemented by NATO. Under this agreement Bosnia-Herzegovina was divided into two parts: the Croat-Muslim Federation and Republika Srpska. The IMF was empowered to appoint a person to run the Bosnian Central Bank, and the European Bank for Reconstruction and Development was instructed to sell off state assets and restructure the public sector. *Newsweek* on 4 December 1995 stated that under the terms of the agreement, US-led NATO forces would 'have nearly colonial powers' in Yugoslavia, thereby giving the US an unprecedented foothold in Europe.

Meanwhile, tension between Serbs and Albanians had been mounting in Kosovo, an area of Serbia with a turbulent past. In Kosovo, ethnic Albanians outnumbered Serbs by nine to one and the province had enjoyed a great deal of autonomy for the last fifteen years. However, in 1989 Serbia had stripped Kosovo of that autonomy. When Serbia joined with Montenegro to declare itself the Federal Republic of Yugoslavia in 1992, Kosovo's desire for autonomy was not addressed. Coupled with this was the fact that as one of Yugoslavia's poorer regions, the province was now suffering from the lack of transfer funds and international aid.

In 1998, Serb leader Slobodan Milosevic met with Richard Hollbrooke, the US special envoy to Yugoslavia who had brokered the Dayton Accord. Milosevic agreed to admit a peacekeeping force to the area but he chose, instead of NATO, the Organization for Security and Cooperation in Europe (OSCE), the official regional security agency under the UN Charter.[6] On 16 October 1998, the Hollbrooke-Milosevic deal was signed, authorising 2000 OSCE representatives to supervise the restoration of order.

Many have blamed the subsequent NATO bombing of Kosovo on the failure of this peacekeeping initiative. However, there were other factors that undermined these attempts to provide a peaceful solution.

The Downward Spiral

Between November 1998 and January 1999, the OSCE sent in a verification force of 200 unarmed, inadequately funded and poorly supported personnel. The number gradually increased to 1200 by March 1999 but never reached the level of 2000 requested. The OSCE representatives reported having some success in conflict resolution and resettlement of internal refugees, but it was fighting a losing battle. Some outside force, not yet properly identified, was providing arms to a rebel Kosovar force, the Kosovo Liberation Army or KLA, which had begun to form as a civilian militia in 1997.

According to Rollie Keith, Canadian member of the OSCE verification team and director of the Kosovo Polje Field Office (just west of Priština), the KLA became fully operative in January 1999 and began provocative attacks on the Serbian security forces. No one is quite sure who armed and trained them. 'This low intensity war since the end of 1998 had resulted in a series of incidents against the security forces, which in turn led to some heavy-handed security operations, one being the alleged massacre at Racak of some 45 Albanian Kosovars in mid-January.'[7] There followed a significant increase in the kidnapping of members of the Serbian security forces and government casualties, which led to major security force reprisals. The OSCE verification forces found themselves being shot at by both sides, and they were often without adequate support or training to handle the escalating conflict.

The OSCE was made up of 54 powerful nations including the US, UK and Russia. It would have been possible to step up the OSCE force from 1200 to 6000 or even 12,000. Sweden, not a member of NATO, had set a positive example by sending personnel who were well-trained in conflict resolution, the language and culture of Serbia, and the history of the conflict, including the names of known leaders in the Kosovan area. Fifty per cent of its delegation went as unarmed military in civilian clothes, 30 per cent were from its trained police force, and 30 per cent were senior personnel from non-governmental organisations. Not all countries were so generous; not all volunteers were as well prepared.

According to renowned peace scholar Johan Galtung, war could have been avoided by focused citizen diplomacy and binoculars, 'living in the villages, and bringing in volunteers to help'.[8] Why this strange failure on the part of such well-positioned nations? Part of the answer must surely lie in the funding of the OSCE – the budget given to NATO by its 19 member states is 1,000 times greater than the budget given by these same states plus 35 more to the OSCE. Galtung went on to say that in the midst of the civil war, after February 1998, US special envoy Robert Gelbard told the Serb government in Belgrade that the US was of the view that the KLA were terrorists; this only served to embolden Milosevic in his suppression of the 'rebel' forces. Instead of exploring further diplomatic solutions, the situation was allowed to deteriorate into outright civil war.[9]

The event that initiated NATO's attack on Serbia, the Racak massacre, was first identified and publicized by William Walker, the US head of the OSCE verification team. Walker said he had found an open grave filled with the bodies of Kosovar civilians massacred by Serbs. It was interpreted as evidence of the beginning of genocide.

Walker was the highly controversial Deputy Chief of Mission in Honduras at the time of the Contra scandal. Unknown to Congress, the CIA had been secretly diverting profits from arms trading with Iran to fund Contra forces in Nicaragua, with the aim of overthrowing the Marxist-orientated Sandinista government. In 1985, Walker was named Assistant US Secretary of State for Central America and special assistant to Elliott Abrams. Elliott Abrams and Lieutenant Colonel Oliver North were the principal members of a Restricted Interagency Group (RIG) that worked on Central American issues for the Reagan administration. Walker frequently accompanied Abrams to the meetings of the RIG and assisted in carrying out the plans it devised. The Independent Counsel Lawrence Walsh's indictment of Abrams and Oliver North in 1988 named Walker as responsible for setting up the phoney humanitarian operation in El Salvador to funnel guns, ammunition and supplies to the Contra rebels. In 1989, when Salvadoran soldiers executed six Jesuit priests, their housekeeper and her 14-year-old daughter, Walker commented at the press conference: 'Management control problems can exist in these kinds of situations.' Although the US never admitted to having more than 50 military advisors in El Salvador during Walker's tenure, in 1996 Walker hosted a ceremony in Washington DC to honour 5000 American soldiers who secretly fought in El Salvador.[10]

This same Walker, who recommended to Secretary of State James Baker that the US not jeopardise its relationship with El Salvador by investigating the deaths of the Jesuits, 'however heinous', called on NATO to wage war over the 'heinous deaths' at Racak.

Both *Le Monde* and *Le Figaro*, respected mainstream French newspapers, as well as French national television, have run articles questioning the Racak incident.[11] These publications cite inconsistencies in Walker's story, the absence of shell casings and blood in the trenches where the bodies were found, and the lack of eyewitnesses despite the presence of journalists and observers in the town during the Serb-KLA fighting. In America, only the *Los Angeles Times* queried this event.[12]

Our understanding of the pre-war situation in Kosovo is further muddied by the fact that there were spy operations going on within the OSCE itself. CARE Canada, which is a private non-governmental organisation (NGO) aid agency, was paid $3 million by a Canadian government agency to recruit and hire OSCE monitors whom it then provided to the Canadian spy agency for training and use in Kosovo prior to NATO bombings. CARE

Canada does not deny that it provided about 60 such spies.

The period of OSCE monitoring was an intense period of 'listening' and 'seeing' for NATO. Spies in Kosovo were able to provide on-the-ground information about troop movements and land mines. It was Malcolm Fraser, former prime minister of Australia and now chair of CARE Australia, who blew the whistle on the Canadian OSCE members, saying that this violation of traditional aid group neutrality endangered all aid workers. Stephen Wallace, an official with the Canadian International Development Agency (CIDA), the agency which funded CARE Canada, said the government frequently uses private agencies to find Canadian staff for international 'peace programmes', a process sometimes called 'mission creep'. Alex Morrison, president of Canada's Pearson Peacekeeping Centre, says: 'At the Pearson Peacekeeping Centre we don't use the term mission creep. We use the term mission reality. By that we mean civilians and military must work together to use resources to the maximum extent.'[13] Expecting personnel to engage in conflict resolution to prevent war, while spying in order to gain military advantage seems a huge contradiction in terms.

Certainly the picture of pre-war Yugoslavia was mixed, and whilst the international disputes about intervention raged on, many peacekeepers were reporting genuine progress. Rollie Keith writes:

> Within our field office area of responsibility, we were making progress to facilitate the resettlement of an unoccupied village from the previous summer, while six other villages were about to be abandoned due to the increasing hostilities. As an example of this humanitarian work, we had conducted some dozen negotiating sessions with both belligerents as well as displaced villagers. Our objective was to create conditions of confidence and stability and commence the resettlement of the village of Donje Grabovac.

Freely admitting that, when violence occurred, it was difficult to tell if it had been provoked by the Serb security force or the KLA, he stated:

The situation was clearly that KLA provocations, as personally witnessed in ambushes of security patrols which inflicted fatal and other casualties, were clear violations of the previous October's agreement. The security forces responded and the consequent security harassment and counter-operations led to an intensified insurrectionary war, but as I have stated elsewhere, I did not witness, nor did I have knowledge of any incidents of so-called 'ethnic cleansing' and there certainly were no occurrences of 'genocidal policies' while I was with the Kosovo verification mission in Kosovo.[14]

The Rambouillet Agreement was presented as a last attempt to deal with Yugoslavian President Slobodan Milosevic. It provided for a very broad form of autonomy for Kosovo, granting the region its own parliament, president, prime minister, supreme court and security forces. All Serb army and police forces would have to be withdrawn, except for a three-mile-wide stretch along the borders of the province. The sticking point was that a 28,000-strong NATO occupation army, known as the KFOR, would be authorised to 'use necessary force to ensure compliance with the Accords' and would be given full diplomatic immunity from all Yugoslavian laws. Although the Serb delegation expressed readiness to accept the political pact at a meeting outside Paris on 15 March 1999, they did not accept military occupation by NATO. The Kosovars signed, but on 23 March the Serbian Parliament rejected the deal. On 24 March, despite international opposition, the NATO bombing began.

Seventy-Nine Days of Horror

Serbia reacted with anger and revenge towards the Albanian Kosovars after they signed the Rambouillet treaty. If both Yugoslavia and Kosovo had rejected it, NATO would not have been able to act. In March 1999, Norwegian Foreign Minister Knut Vollebaeck, who was both the Chair in Office of the OSCE and the Norwegian representative on the governing board of NATO, ordered the evacuation of the OSCE in preparation for the NATO bombings. With the unleashing of violence came the ethnic cleansing and genocidal attacks on civilians, refugee migration and all the terrible consequences of anger, hatred, fear and war.

In its reporting of the war, the Western media was selective. Many of NATO's mistakes were highly publicised, like the bombing of the Chinese Embassy or of civilian buses, but many other disasters were neglected – the petrochemical clouds released by the bombing which poisoned agricultural products, the only food of the ordinary people, and the oil slick in the River Danube which threatened not only drinking and irrigation water, but also the water intake of the nuclear power reactor near Belgrade. The bridges that were blown up in order to reduce military mobility also cut off food supplies to civilians; the disruptions in the electricity supply affected incubators and other life-saving equipment in the region's hospitals.

Some of the worst devastation was wrought by NATO's use of depleted uranium (DU) bombs and missiles.[15] However, the war was almost over before the US admitted to having used these weapons. This US radioactive waste has been given free of charge to military manufacturers and is used in place of tungsten and lead in missiles and bullets. It leaves a legacy of radioactive debris and contamination on a battlefield and was used extensively for the first time in the Gulf War.

The story of depleted uranium weapons can be traced back as far as 1943, or even earlier if one considers the suffering of the first uranium miners in Czechoslovakia. In a document dated 30 October 1943 and labelled 'War Department, United States Engineer Office Manhattan District' (which was the code for the Manhattan project which produced the atomic bomb), we find a remarkable list on the use of radioactive materials as a military weapon. (This document was declassified in 1974.)

> The material was to be ground into particles of microscopic size to form dust and smoke and distributed by a ground-fired projectile, land vehicles or aerial bombs. In this form it would be inhaled by personnel. The amount necessary to cause death to a person inhaling the material is extremely small. It has been estimated that one millionth of a gram accumulating in a person's body would be fatal. There are no known methods of treatment for such a casualty. Two factors appear to increase the

effectiveness of radioactive dust or smoke as a weapon. These are, it cannot be detected by the senses, and it can be distributed in a dust of smoke so finely powdered that it will permeate a standard gas mask filter in quantities large enough to be extremely dangerous.

Among the possible uses for such weapons was 'the ability to use it against large cities, to promote panic and create casualties among civilian populations'. The document called for the immediate formation of a research group at the University of Chicago. Would a young, eager student putting his mind to this 'problem' have had any idea that fifty years later US soldiers would be returning from the Gulf War contaminated and ill?

When fired, depleted uranium burns at over 3000 degrees centigrade (like the firing of pottery in a kiln) and becomes a ceramic uranium aerosol containing microscopic radioactive particles. These particles can travel far from the point of release and can be inhaled. They stay in the body for many years, irradiating all of the tissues and organs near to where they lodge. They also affect the environment and wildlife. In a study at Elgin Air Force Base, Wayne Hanson and Felix Miera tested soil at various distances from armour plate targets which had been struck by 'depleted uranium penetrators'. Vegetation samples contained 320 parts per million (ppm) of uranium in 1974. A year later they still contained 125 ppm. Small mammals trapped in the study areas contained 'a maximum of 210 ppm of uranium in the gastro-intestinal tract contents, 24 ppm in the pelt and 4 ppm in the remaining carcass'. In June of the next year maximum concentrations were '110, 50 and 2 ppm respectively, with 6 ppm in the lungs' emphasising 'the importance of resuspension of respirable particles in the upper few millimetres of the soil as a contamination mechanism'.[16] Like landmines, these weapons continue to cause harm long after the end of the war.

Even the production and testing of DU weapons can be problematic. Following concerns about excess childhood leukemia rates in the vicinity of AWE Aldermaston, where British nuclear and DU weapons are produced and stored, the base was required to monitor for uranium dust at its own perimeter fence and in many sites in West Berkshire and North Hampshire. Studies found that levels of uranium dust are even higher near military testing sites than at the Aldermaston base.[17] The air concentration of uranium experienced by personnel entering a firing range one hour after testing was 45,000 times higher than the figure on-site at Aldermaston. This level exceeds the limits permitted by the National Radiation Protection Board by a factor of 6.75, and is illegal in the UK.[18]

The United Nations Human Rights Tribunal adopted resolutions in 1996 and 1997 which included depleted uranium weapons among 'weapons of mass and indiscriminate destruction' incompatible with international humanitarian or human rights law.[19] The question of their use is now being debated both within the United Nations and in the International Court of Law. So-called 'precision bombing' is a cruel hoax when DU is used because its dispersion is uncontrollable. The Human Rights Tribunal appointed Clemencia Ferero de Castellanos, a Colombian delegate, as rapporteur of a follow-up study on DU and other weapons of mass and indiscriminate destruction. It appears that she has so far not published her brief. Since she has recently been appointed to a government position in Colombia, critics are suggesting that it may now be appropriate for her to turn the task over to another rapporteur.

After the 79 days of bombing in Kosovo, refugees wanted nothing more than to return to their homes. Although the British Ministry of Defence (MoD) warned its own personnel about the possibility of DU contamination, it failed to warn the refugees:

> Ministry of Defence personnel in Kosovo have been warned to stay clear of areas which have been affected by depleted uranium weapons unless they are wearing full radiological protective clothing. However, returning refugees have been kept in the dark about the perils of moving back to the highly contaminated areas, with the MoD claiming that responsibility for alerting them lies with United Nations relief workers. When asked if there was a co-ordinated NATO response relating to the returning refugees, the locals rebuilding, and to the advice to avoid disturbing areas of depleted uranium (DU) contamination, an MOD spokesperson

replied: 'There's no specially reviewed policy re DU. It would have to be co-ordinated by NATO. We would follow and adhere to any of their directions.'[20]

Assessing the Damage

Through the intervention of Mikhail Gorbachev and the International Green Cross, the Kosovo conflict was the first to be followed by an international inspection of the environment by a UN agency, the United Nations Environment Program or UNEP. Two UN agencies, UNEP and the UN Committee on Human Settlements, sent a delegation to Serbia and Montenegro in June 1999. On their return, they called a conference including both governmental and non-governmental organizations, and the Balkan Task Force of Environment and Human Settlements (BTF) was formed. Greenpeace, the World Wide Fund for Nature, the Green Cross and the Danube River Commission joined the BTF. The team was under the leadership of UN Under Secretary General, Sergio De Mello. The *Boston Globe* reported on this mission on 6 August 1999:

> NATO's bombing of Yugoslav industrial sites has contaminated the river Danube and ground water in parts of Serbia and Kosovo, posing a health hazard for several years. 'We have found that on many of these targeted sites there are serious environmental consequences and probably also serious health consequences.'[21]

According to UN News Release No. 70 1999:

> A large number of civilian industrial facilities (more than 80) have been attacked and destroyed in the NATO air campaign. Damage to oil refineries, fuel dumps and chemical and fertilizer factories, as well as the toxic smoke from huge fires and leakage of harmful chemicals into the soil and the water table, have contributed to, as yet un-assessed, levels of environmental pollution in some urban areas, which will in turn have a negative impact on health and ecological systems.

The mission visited Pancevo, 15 kilometres northeast of Belgrade, where the destruction of a petrochemical plant resulted in the release of various chemical fluids (such as vinyl-chloride, chlorine, ethylene-dichloride, propylene) into the atmosphere, water and soil. Many of these compounds can cause cancer, miscarriages and birth defects. Others are associated with fatal nerve and liver diseases.

> The pollutants which have been released could also have a negative effect in the short and long term on the nutrition chain. The lack of protective substances, as well as fertilizer, could also endanger the survival of certain plants. Land, rivers, lakes and underground waters may be polluted due to the spillage of petrochemicals, oil spills and other chemicals. The ability of the local authorities to undertake decontamination and recovery in an environmentally sound manner is hampered by shortages of material and equipment.[22]

It will be many years before the full impact of the destruction in Yugoslavia and Kosovo is known. When the final report of the environmental inspection was released, NATO issued information on its use of DU for the first time. In a letter dated 7 February 2000, NATO Secretary General Lord Robertson confirmed to UN Secretary General Kofi Annan the following:

> A total of approximately 31,000 rounds of DU ammunition was used in operation Allied Force. The major focus of these operations was in an area west of the Pec-Dakovica-Prizren highway; in the area surrounding Klina; in the area around Prizren; and in an area to the north of a line joining Suva Reka and Urosevac. However, many missions using DU also took place outside of these areas. At this moment it is impossible to state accurately every location where DU ammunition was used. Attached is a map providing the best available information as to the location of DU ammunition use.

The Investigative Committee had needed this map for its mission so that it could sample the environmental residue of DU. No sampling was undertaken, however, because NATO was so late with supplying the information.[23]

The Legal Implications

In the aftermath of the bombing, the visible destruction caused by both Serb and NATO forces dominated the debate; the subtle long-term consequences of contamination of land, air and water escaped notice. On the popular level, constant television coverage focused on the 'humanitarian dimensions', especially the pitiful conditions of the refugee camps. That people were suffering is not in dispute, but NATO wanted to impress on the public the idea that it was acting to enforce humanitarian law. Ironically, according to an article by Richard Norton-Taylor in the *Guardian Weekly* (13 January 2000), Human Rights Watch, a New York-based NGO, accused NATO of deliberately bombing Serbia's civil infrastructure and using cluster bombs, in direct violation of humanitarian law.

The bombing of Kosovo highlighted many serious questions about the mechanisms and enforcement of international law and the boundaries of jurisdiction. As noted earlier, NATO had acted without specific authorisation from the UN. The US and other NATO countries were able to convince the world that they could not take the Yugoslavian question to the UN Security Council because Russia, and possibly China, would block any action with a veto. This was not the only option. The US, UK and Canada, which are NATO countries, together with Russia, Ireland, Sweden, and Switzerland, which are non-NATO countries, are all members of the OSCE. There is no veto power in the OSCE. Moreover in 1950, in order to lessen the veto power of the five nuclear nations, the UN provided that whenever the veto was used, the Secretary General could call an emergency meeting of the full General Assembly within 48 hours. Therefore NATO had at least two other options for seeking approval of its course of action.

In an attempt to stop the bombing, the Federal Republic of Yugoslavia had appealed to the World Court for an injunction. I spent a day at the court during the response to cases filed separately against twelve of the NATO nations. Most of these countries claimed that the Federal Republic of Yugoslavia was not the same country that had been admitted to membership of the United Nations, and therefore it had no standing before the court. When the court finally ruled on 2 June 1999, it declared that it did not have jurisdiction enabling it to handle the case. What I found ironic was that the UN membership of Yugoslavia, which had paid its UN dues, was being questioned by the United States, which was seriously in arrears. Moreover, the UN had tacitly accepted the Federal Republic of Yugoslavia as the lawful successor of Yugoslavia when it accepted its dues. Its status as a member of the UN was therefore not in question.

A fuller consideration of the question of jurisdiction will take place later. The Court accordingly remains seized of those cases and has reserved the subsequent procedure for further decision. In its reasoning, the Court expresses its deep concern 'with the human tragedy, the loss of life, and the enormous suffering in Kosovo which form the background' of the dispute and 'with the continuing loss of life and human suffering in all parts of Yugoslavia'. It sets out its profound concern with the use of force in Yugoslavia, which 'under the present circumstances...raises very serious issues of international law', and emphasizes that 'all parties before it must act in conformity with their obligations under the United Nations Charter and other rules of international law, including humanitarian law'.[24]

When NATO granted itself the power to condemn, judge and punish Yugoslavia, it set a dangerous precedent for lawlessness and vengeance in the international arena. This danger is compounded when the World Court finds itself not empowered to act and the United Nations is considered useless. The barbarous revenge which Yugoslavia unleashed against the Albanians in Kosovo had no justification, of course, and it shocked the world, but I hope this is enough to convince the reader that there is 'another side' to the conflict. The war was widely covered in the media as necessary humanitarian intervention. This coverage suppressed the war's political and economic goals and the dangers caused by NATO's own use of high-tech weaponry. At each step along the road to war there were viable alternatives; but NATO successfully convinced the public that all options had been tried, and they had failed.

Many of NATO's actions were formally submitted to the War Crimes Tribunal at the Hague, along with the crimes against humanity which are being attributed to Serbia. As expected, on 2 June 2000 Carla del Ponte, the war crimes prosecutor, told the UN Security Council that there was 'no basis' for opening a trial against NATO personnel, as she was satisfied that although NATO had made some mistakes, there was no deliberate targeting of civilians or unlawful military targets during the campaign.[25]

THE PRIME-TIME WAR: IRAQ

One evening as I was going home from work, I overheard some young people talking about the Gulf War. It was in March of 1991, and the war had started the previous January. These students were talking about going home to watch the 'prime-time war'.

The overt military build-up began after the Iraqi invasion of Kuwait on 2 August 1990, yet the tensions that produced it could be traced right back to the 1930s, when Great Britain, France and the United States dominated the Arabian Peninsula and its oil reserves. The major players in the trade were five American companies, Exxon, Mobil, Chevron, Texaco and Gulf, plus Royal Dutch Shell and the Anglo-Persian Oil Company (now BP). These were dubbed the 'Seven Sisters' and they controlled refineries, pipelines, tankers and crude oil production throughout the world.

The links between politics, oil and war have always been strong. When Iran nationalised its oil industry in 1951, the US and British governments supported a coup to overthrow the prime minister and replace him with the Shah. Iran then bought billions of dollars' worth of US weaponry and became a major regional distributor of US products. In 1972 when the Iraqi government nationalised the Iraqi Petroleum Company, President Nixon began arming the Kurdish people in an attempt to destabilise the regime. Then when Iraq agreed to share the Shatt-al-Arab waterway (a vital part of trade infrastructure) with Iran in 1975, the US support for the Kurds was abruptly terminated. After the fall of the Shah in 1979 and the accession of Ayatollah Ruholla Khomeini to power, the US became supportive of Iraq and did not complain when Iraq attacked Iran in 1980, beginning the eight-year Iran–Iraq war. Following these events and the terrorist seizure of the US Embassy in Teheran, the US, Russia, Saudi Arabia, Kuwait and most of the other Emirates provided military aid and assistance to Iraq. To complicate matters, in 1985 and 1986, the US also secretly supplied Iran with arms in what became known as the Iran-Contra scandal (see p16). When the war between Iran and Iraq ended, the propaganda against Saddam Hussein began.

During the 75 years of Western dominance in the Gulf, the US and UK had never shown great concern for democratic values, human rights, or social justice. When it suited their purpose, they supported Iraq in its violent aggression against Iran, ignoring Iraq's poor human rights record and its violent attacks on its own people.[26] The about face of US policy in 1990 and its mobilisation of countries against Iraq was obviously based on domestic oil needs. US oil production had declined during the 1980s. Oil experts predicted that US oil imports from the Gulf, which had been 5 per cent in 1973, would rise from 10 per cent in 1989 to 25 per cent by the year 2000.[27] The dependency of Europe and Japan on Gulf oil was even greater.

Perhaps there was also a desire to test the newest military doctrine known as air/land battle. As a 'logistics exercise', the Gulf War has been described as the equivalent of a second Normandy invasion, funnelled into 40 days of actual combat.

During the 1980s Saudi Arabia poured almost $50 billion into creating a Gulf-wide air defence system built according to NATO specifications. By 1988, the US Army Corps of Engineers had also constructed a $14 billion network of military cantonments across Saudi Arabia. These buildings were being expanded to accommodate a large troop deployment even as the Iraqis massed on Kuwait's border ready to attack.[28] Early in 1990, before Iraq invaded Kuwait, General Schwarzkopf informed the US Senate Armed Services Committee of the new military strategy designed to protect US access to and control over Gulf oil in case of regional conflict. This technique, involving a kind of high-speed blitzkrieg, had been designed during the Cold War period for use against Soviet tanks pouring into central Europe. The strategy included use of depleted uranium ordnance on a large scale for the first time.[29]

It is now a matter of record that prior to the Gulf War, the US Congress had approved agricultural loan subsidies of hundreds of millions of dollars to Iraq. This loan would benefit US farmers in that the Iraqis would buy rice, corn, wheat, and other essentials almost exclusively from the US. These loans were suddenly cut off in an economic embargo imposed by President Bush.

The imposition of sanctions on basics such as food always hits the civilian population first, and it hits them hard. The US blockade on food sales caused serious shortages in Iraq, provoking internal unrest. However, US manufacturers' sales of arms to Iraq were not cut off immediately. On 27 October 1992, the House of Representatives learned that US corporations, with government agency consent, as well as corporations in other NATO countries, had exported chemical, biological, nuclear and missile system components to Iraq as late as 1989. The amounts from the US were significant – for example, 1500 gallons of anthrax and 39 tons of biological warfare agents. The latter were produced in Boca Raton, Florida; items for scud missiles were made in Connecticut and Pittsburgh. Even more astonishing, when Iraq failed to pay for some of this military ordnance, the American taxpayer had to foot the bill.[30] Russia, France, UK and Germany also supplied Iraq with weapons during the 1970s and 1980s in spite of the fact that Iraq was thought to have the fourth largest army in the world, posing a credible threat to its neighbours.[31]

Arms sales make money, but they don't necessarily give the purchaser military capability. When a country sells a weapons system to another, it then can get funding from its own government to create a counter-weapon. This keeps the military machine well-oiled. The Iraqi army was large, but the military hardware sold to them by the West did not pose a threat to NATO countries as they had already developed the capability to defend themselves. They were, however, a considerable problem for the other countries surrounding Iraq and for the Kurdish minority.

The Build-up to War

Kuwait wasn't really surprised by the Iraqi invasion. Iraq had appealed to both Kuwait and Saudi Arabia for release from foreign war loans, and an agreement was drawn up which would address both this issue and a longstanding border dispute. The treaty was due to be signed in Jeddah on 3 August 1990. When Jordan's King

Hussein urged Kuwait to get on with this agreement, the Kuwaiti Foreign Minister replied: 'We are not going to respond to Iraq. If they don't like it, let them occupy our territory. We are going to bring in the Americans.'[32]

According to the *Guardian Weekly* (13 January 1991), which obtained an Iraqi transcript of the meeting, US Ambassador April Gilespie met with Saddam Hussein on 25 July, a week before the invasion of Kuwait. The US State Department has not denied the accuracy of this document. When Saddam Hussein stated that 'Iraq could not accept death in the face of Kuwait's economic war and military action', she failed to warn him that the US would oppose an Iraqi invasion. She explained later that she had 'direct orders from the President to seek better relations with Iraq'.[33] Four days before the invasion, the US Senate reported that the CIA were predicting the invasion to the day. Two days later, the Assistant US Secretary of State John Kelly told a Congressional hearing that the US was not committed to defending Kuwait. No action to stave off attack was taken.[34]

The Iraqi troops not only invaded Kuwait on 2 August 1990, but they went all the way to the capital city and took control of the whole country.

The next move was to gain support from the surrounding Arab nations. On 3 August, President Bush sent Secretary Cheney, General Powell and General Schwarzkopf to Saudi Arabia. They told King Fahd that the US believed Saddam Hussein could attack Saudi Arabia in as little as 48 hours, thereby destroying any prospects of an Arab resolution to the problem. President Bush promptly ordered 40,000 US military personnel to Saudi Arabia, and subsequently added another 200,000, although he waited until after the November election to notify Congress. In October 1990, General Powell referred to the new military plan for waging war against Iraq that had been readied for such an emergency.[35]

According to former US Attorney General Ramsey Clark, the United States then manoeuvred the UN Security Council into an unprecedented series of resolutions against Iraq, eventually securing one, on 29 November, which authorised any nation to use 'all necessary means to enforce these resolutions':

The US paid multi-billions of dollars in bribes, offered arms for regional wars, threatened and carried out economic retaliation, forgave multi-billion dollar loans, offered diplomatic relations despite human rights violations, and in other ways corruptly exacted votes, creating the appearance of near universal international approval of US policies toward Iraq.[36]

Iraq was to withdraw from Kuwait by 15 January or face the consequences.

Propaganda is a vital part of any war, and the next question was how to win the support of the American and European people. An anonymous American organisation called 'Citizens for a Free Kuwait' hired the prestigious public relations firm Hill Knowlton to develop programmes which would generate public support for a UN attack on Iraq. According to Hill Knowlton, 'President Bush was kept informed of what was being done'. It was this firm that publicised the story, distributed worldwide, that Iraqi soldiers had thrown babies out of incubators in a brutal takeover of Kuwaiti hospitals.[37] The story was later discredited when it was discovered that its star witness was the young daughter of the Kuwaiti Ambassador to the United States.

Meanwhile, the British Minister of Defence briefed journalists on what was 'non-releasable information' for Gulf War reporters: no numbers of troops, aircraft or other equipment. No names of military installations or specific location of military units. No information on future operations or security precautions. No details on the rules of engagement, or information on intelligence collection. Basically, the media was to be provided with 'safe' information through special government and military briefings.[38]

The Reality on the Ground in Iraq

The Gulf War provided a first venture for NATO forces outside the European theatre, and it included not only the US, Britain, France and Canada, but also a rear guard of almost 150,000 troops from Turkey with the help of about a dozen Arab nations.[39] Belgium, Denmark, Germany, Greece, Italy, Netherlands, Portugal, and

Spain each sent a few ships and aircraft. Arranging such an alliance was considered to be a monumental foreign policy achievement for President Bush.

Operation Desert Storm

Systematic carpet bombing of Iraq began on 16 January 1991 at 6.30 pm Eastern Standard Time (2.30 am in Baghdad) – right on time for the US prime-time news.

The Gulf War provided the first battle-testing of dozens of highly sophisticated weapons, and a fully integrated electronic battlefield. The 'accuracy' of these smart weapons was demonstrated regularly for the evening television viewer. Pictures such as those of a laser targeting system lining up on the roof of the Iraqi Ministry of Defence, and a 2000-pound laser bomb blowing the building apart, gave the impression of a remote, bloodless, push-button war. Some of this 'magic' was dispelled when both the *New York Times* and *Boston Globe* attacked the 'surgical myth of the Gulf War' on 16 April 1991. Apparently only 7.4 per cent of the bombs used were so-called 'precision-guided ordnance'.

There were more than 110,000 air sorties (a rate of nearly two per minute over a period of six weeks), and more than 88,000 tons of explosives were dropped. The destruction is said to have been equivalent to seven times the force of the Hiroshima bomb. After the war, United Nations investigators described civilian damage as 'near Apocalyptic'.

The planes met very little resistance from Iraqi aircraft, and Iraq had no effective anti-aircraft or anti-missile defence. The Iraqi scud missiles, which were talked about so much, were surface-to-surface missiles designed for long-range penetration into 'enemy' country – Saudi Arabia or Israel. The main targets of the NATO bombing were electrical power generation, relay and transmission; water treatment plants, together with pumps, distribution systems and reservoirs; telephone and radio transmission, and relay facilities; food processing, storage and distribution warehouses and markets; beverage plants, including infant milk formula; animal vaccination stations; agricultural irrigation sites; railroads; bus depots; bridges, highway overpasses; public transportation vehicles; oil wells, pipelines, oil storage depots, gasoline filling stations, kerosene storage tanks; sewage treatment and disposal systems; and factories engaged in civilian production (for example, cars and textiles). It has been estimated that thousands of civilians died from dehydration, dysentery and disease caused by contaminated water and inability to receive medical assistance. Even more died from hunger, shock, cold and stress due to lack of food, sanitary living conditions, housing and the other necessities of life.

Ramsey Clark, travelling more than 2000 miles through Iraq with a camera crew on 2–8 February 1991, reported extensive damage: 'No city, town or roadside stop had any running water, electricity, telephone or adequate gasoline for transportation.'[40] His report matched those of others who went to Iraq after the war ended. Dr Ibrahim Al Moore, who was head of the Red Crescent for ten years and delivered Red Cross medicines to civilian hospitals in Iraq, estimated that by 6 February 1991, civilian deaths from the bombing were 6,000–7,000, with another 6,000 dead from contaminated water, lack of medicine, and insufficient supplies of infant formula.[41]

It is also thought that chemical and perhaps biological warfare factories in Iraq were struck by bombs. Obviously destruction of such highly toxic facilities merely spreads this hazardous material into the surrounding environment, posing a threat to humans and animals. It was thought that Iraq had a stockpile of thousands of tons of mustard gas, a blister agent used in World War I. This gas is persistent in the environment, and in France has been reported to be burning people 70 years after it was originally used. Chemical and biological brews, indiscriminately disseminated by a bomb, could send toxic clouds over highly populated areas – including over Allied personnel.

Even more problematic was the bombing of Iraq's two nuclear research reactors some time after 17 January 1991. These two reactors were located in the southern suburbs of Baghdad. They included a Russian-supplied five-megawatt (thermal) reactor – commercial nuclear reactors are about 1000- to 3000-megawatt thermal – and an even smaller Tammuz-2, 0.5-megawatt reactor supplied by the French. Iraq had originally had three nuclear reactors. The largest, located at the Osirak Complex, was destroyed by Israel in a 'pre-emptive strike' in 1981, before its nuclear fuel was loaded. The two smaller reactors were operative and produced some heat, electricity and nuclear tracers for medical use. Reportedly, high-level radioactive waste was also stored on site. According to the US military: 'precision bombing caused the reactors to collapse in on themselves – sealing in the nuclear fuel and fission by-products under tons of rubble.'[42] One wonders if this story of perfect containment is more wishful thinking than fact.

Although many Americans and Europeans thought that Iraq was being attacked to prevent it developing nuclear weapons, Iraq had no such capability. These two research reactors used a total of 6 kilograms of fuel enriched to 80 per cent in uranium 235, insufficient for making uranium bombs, which require a minimum of 22 kilograms of fuel enriched to at least 95 per cent. It is also inappropriate for making plutonium bombs because one needs uranium 238 to generate the plutonium. We do now know that Iraq was trying to build a uranium enrichment plant, and may eventually have been able to enrich uranium itself without having to depend on Russia or France for supplies, but this would have required many more years. During 'Safeguard Inspections' in November 1990, the International Atomic Energy Agency declared that Iraq had accounted for all of its nuclear fuel. Iraq was also a signatory to and in compliance with the Nuclear Non-Proliferation Treaty.

Operation Desert Sabre

The tank war began on 24 February, directly following the heavy carpet bombing of Iraqi positions and the most massive helicopter assault since Vietnam. US AH-64 Apache 'tank killing' helicopters directly preceded ground attacks from both east and west. The Iraqi arsenal consisted mostly of Soviet tanks, designed 20 years ago, with a main gun range about 1000 metres short of the range of Allied tank guns.[43] Even when their range was close enough to hit a NATO Bradley tank, it had little effect, as the Bradley had a hardened depleted uranium shell to protect it.

The Iraqis did not possess DU weapons, but there were fears that they might use chemical warfare against the Allied forces. The

British MoD believed that Iraq 'may have as many as 100,000 artillery shells, filled with chemicals and several tons were stored near the front line'.[44] It was a rational fear, since it was the West who exported this ordnance to Iraq.

According to a first-hand report, 30,000 bomb-stunned and starving Iraqis surrendered in the first two days. Many of the conscripts were as young as 11, 12 and 13 years old, and they had had just six weeks of military training before being sent to war. At least 40 Iraqis tried to surrender to a CNN television crew. The Iraqis were exhausted, hungry and in a very distressed condition. Most seemed relieved to surrender.

Along what has been called the 'Highway to Hell', where Allied forces pounded the retreating Iraqi conscripts, there were more than 2,000 wrecked vehicles and 10,000 to 15,000 charred bodies. Disputes continue to rage about an attack two days after the ceasefire, and it was the subject of a 1991 investigation by the US Army's Criminal Investigation Command. The military investigators, fielding an anonymous complaint and completing a secret report, exonerated Barry R. McCaffery, commander of the Army's 24th Infantry Division, which had continued destroying Iraqi positions after the cease-fire was announced. After his exoneration, McCaffery was made a four-star general, retired, and was then appointed President Clinton's top drugs control official. The incident has been revived by an investigative reporter, Seymour M. Hersh, and questions raised about the investigation's conclusions in a long article based on some 200 interviews (including members of the 24th Infantry) in *The New Yorker* magazine.[45]

Army engineer Dwayne Mower identified DU hits on nearly half of the thousands of buses, trucks, cars and tanks he saw on the highway. At the time, he thought that talk of radioactivity was just a rumour, so he and his comrades did not worry when a 40-ton transport truck crammed with DU rounds accidentally blew up near their camp. Mower and most of the 651st Combat Support Attachment later began experiencing strange flu-like symptoms.[46]

Along the highway, the effects were much, much worse. On some bodies the hair and clothes were burned off, and skin was incinerated by heat so intense that it melted the windshields of vehicles onto the dashboards. Napalm, cluster and anti-personnel fragmentation bombs were used. Bomblets, called Sadeyes, which can either explode on impact or be timed for later detonation, were carried in missiles capable of spreading deadly shrapnel over an area equal to 157 football fields.[47]

A secret document from the offices of the Atomic Energy authority smuggled out to a *Guardian* reporter stated that the depleted uranium shells used by the Allies in the Gulf War had left 'at least 40 tons of radioactive dust on the battlefields of Kuwait and Iraq.'[48] Maps from the Pentagon later showed most of southern Iraq and northern Kuwait covered with depleted uranium fallout. During a major fire at the Doha Base used by the US Army in Kuwait, there were six hours of severe explosions and raging fires, releasing sufficient aerosols to engulf much of the battlefield. There was a steady eight-knot wind, sending the smoke and debris to the south south-east.[49] Estimates are that hundreds of tons of uranium were left on the battlefield, on the destroyed tanks and vehicles. Worse still, some of the contaminated shells were taken home as souvenirs. The US has stated that it had no legal obligation to clean up the battlefield, and reinforcing this position has commented: 'It does not appear that Kuwait has addressed the long-term management of hazardous and radioactive materials in captured vehicles.'[50]

By the end of the war, the Allies had suffered damage to eleven tanks: four were struck by land mines and seven were hit by cannon fire. None of the damage was serious. There were approximately 300 Allied deaths in the war, and some were due to so-called 'friendly fire', including a rather incredible incident in which US forces shot at a British unit.[51]

The official estimate of damage to the Iraqi side was 100,000 troops killed, 85,000 captured and 100,000 deserted, with over 300,000 wounded. The actual number of Iraqis killed has never been determined. Under the Geneva Conventions, armies are required to maintain a grave registration service and exchange 'lists' after the war:

> parties to conflict shall ensure that burial or cremation of the dead, carried out individually as far as circumstances permit, is

preceded by a careful examination of the bodies with a view to confirming death, establishing identity and enabling report to be made ... they shall further ensure that the dead are honorably interred, if possible according to the rite of the religion to which they belonged, that their graves are respected ... properly maintained and marked so that they may always be found.

In the aftermath of the Gulf War, NATO failed to provide the Red Cross with the names of the tens of thousands killed, or the locations of mass graves – in fact, the Red Cross only gained access to graves for 21 people. This has been one of the most disturbing mysteries of the Gulf War and is in direct violation of the 1949 Geneva Conventions.[52] No one can estimate the number of deaths and illnesses that will result from the long-term consequences.

Aftermath of the 40-Day War

One of the first foreigners to visit Baghdad after this short war was Rich McCutcheon, coordinator of the Canadian Friends Service Committee. He travelled with a Red Crescent medical convoy, which arrived in Iraq on 24 March 1991, bringing one and a half tons of baby milk powder, one and a half tons of intravenous fluid, and about $20,000 worth of medicines, including drugs for diabetes, asthma in children and heart disease.[53]

McCutcheon reported that the convoy found most roads impassable. With all means of communication destroyed, water purification systems taken out, sewage systems disabled, and public transportation obstructed, the entire city of Baghdad was falling apart. Fuel, food and even doctors had been rationed. It was even worse in Karbala, a holy city 88 kilometres southwest of Baghdad. There, McCutcheon witnessed massive destruction: the top two floors of the hospital had caved in; X-ray machines and refrigerators for keeping blood cool had been destroyed. The Allied bombing had spawned internal violence and unrest. There was house-to-house fighting and looting; stores and homes were burned to the ground. In the hospital in Karbala, physicians, medical staff and patients had been expelled by anti-government forces, who used the hospital as a fortress. Patients who could walk, ran, and those who couldn't were shot. Beds were thrown into the street, and ambulances were systematically and irreparably destroyed.

In 1991, the World Health Organisation and UNICEF sent a mission to assess humanitarian needs in Iraq. The leader of the mission, Under Secretary General Martti Ahtisaari of Save the Children, reported that people were using water from roadside ditches for food preparation and drinking. This was the same water in which children and animals swam, and clothes were washed. There was no fuel or electricity available to boil the water, and no water purification tablets. As summer approached and the days became hotter, children became thirstier. As they drank the polluted water, they got diarrhoea and became more dehydrated.

Sister Ann Montgomery, of the Congregation of Sisters of the Sacred Heart, who lives at the Aletheia School of Prayer in New York City, is well known for her long resistance to US nuclear war policy. In July 1991, she travelled to Baghdad: 'Children in Baghdad beg one piece of bread at each house, hoping to fill a sack. One of the seven or eight children's hospitals there lost 300 children during the bombing. They went for 40 days without food, water or electricity.'[54]

She too visited the hospital in Karbala. Four months after Rich McCutcheon's visit, there were still no sanitary supplies or drugs; very little gauze or anaesthesia for surgery; insufficient supplies of insulin, vaccines, intravenous fluids, and blood plasma. Electricity was available only intermittently. No electricity meant no laboratories, no blood banks, no media culturing, no refrigeration, no sterilisation of equipment and no X-rays. All vaccines were destroyed because they could not be properly stored. There was no haemodialysis for kidney patients, no oral rehydration salts or baby milk formula.

Although the city of Mosul had no military targets, its church, school and poorest neighbourhoods had been bombed, block after block. Almost every family had lost at least one member. The day before Ann Montgomery arrived, four skeletons were dug out of the Syrian Catholic elementary school where a family from Baghdad had taken refuge.

Iraq became a country of famine and food riots, the population bombed into pre-industrialised civilisation in 40 days. Unemployment rose to 70–90 per cent, while the cost of living rocketed. There was an increase in crime, family breakdown, and psychiatric illness. There was also a severe increase in malaria, diarrhoea, gastroenteritis, meningitis, and hepatitis A. According to Ann Montgomery's article, some outbreaks of cholera and typhoid were reported. The destruction of the country's infrastructure was further compounded by the sweeping sanctions imposed by the UN after the end of the war.

In September 1991 Eric Hoskins, a Canadian physician, coordinated an international study team funded principally by UNICEF, the MacArthur Foundation, the John Merck Fund and Oxfam-UK. The team visited Iraq's thirty largest cities and rural areas throughout the region. Separate in-depth studies on child mortality and nutrition, health facilities, electrical facilities, water and waste systems, environment and agriculture, income and economics, child psychology, and women were carried out. The team discovered that, although the war was over, the killing of the Iraqis was still going on. Child mortality rates had risen by about 380 per cent. It should be remembered that in Iraq, before the war, 45 per cent of the population was below 15 years of age, so the sanctions primarily affected children. 'Civilized societies cannot pursue Saddam Hussein by withholding insulin from teenage diabetics,' was Hoskins' comment. Another physician who travelled to Iraq from California was David Levinson, who reported that 'all of the parameters for severe epidemics exist in Iraq: poor sanitation, no communication, lack of food, lack of medicine, lack of transportation and a poor water supply.'[55] It was these physicians who first pointed out the radioactive debris left on the battlefield, and that children were playing with depleted uranium bullets.

On his return to Canada, Hoskins tried to convince the Canadian government to release $2 million of Iraqi funds, which had been frozen since the war, to pay for baby milk and medicine. Other countries were holding Iraqi foreign deposits worth $4 billion. None of this was released as aid. About 1800 tons of milk purchased by Iraq was held in Turkey while the children of Iraq were dying.[56]

Again in 1995, the world was told of the country's suffering. The Washington Report on Middle East Affairs stated: 'health officials have reported alarmingly high increases in rare and unknown diseases, primarily in children. Anencephaly, leukaemia, carcinoma and cancers of the lung and digestive system have risen dramatically, as have late-term miscarriages and incidence of congenital diseases and deformities in fetuses.'[57]

These findings were verified by UNICEF in an Associated Press and Reuters News Service release in 1997 – 'between August 1990 and August 1997, more than 1.2 million children in Iraq had died due to embargo related causes'. The embargo forced surgery without anaesthetics, antibiotics or painkillers. It prohibited medical journals. There was no camera film, and families reported not having any pictures of their children, who were now dead. Toys, bikes, pencils, erasers, and children's exercise books were also prohibited.[58] Prior to the Gulf War, Iraq had enjoyed one of the best health-care systems in the region. Leukaemia had been cured at a rate of 76 per cent; after the war it was only 25 per cent. Even in 1999 the death rate of children and infants was still more than double that preceding the war.[59]

The Wider Impact of the War

The fertile delta between the Tigris and Euphrates rivers is known as the 'cradle of civilisation' and the birthplace of agriculture. The first irrigation systems were developed here seven thousand years ago, and the area hosted the first 'urban settlements'. It is a region rich in archaeological treasures and holy places. Before the war, agriculture in the 'fertile crescent' employed 23 per cent of Iraq's population, and the area produced some of the world's largest date, rice, wheat, barley, fruit, vegetable, and fodder crops.

Iraq also has a fragile desert area, home to many small animals. The desert soils are held together by a living crust of microorganisms, ephemeral plants, salt, silt and sand. We know from experience how long it takes for such systems to recover from tank warfare – during World War II , desert warfare in North Africa caused a ten-fold increase in dust storms.

But for the West, the most 'valuable' of Iraq's assets was, of course, its crude oil reserves. Ironically these reserves caused some of the worst environmental destruction, as oil was spilled onto the land and into the sea, and huge oil fires filled the skies with dense, toxic smoke.

Oil-soaked cormorants staggered ashore on the Saudi Arabian coast after 63 million gallons of crude oil were released in a series of spills.[60] The Socotra cormorants breed only in the Gulf and they feed by diving for fish. They became coated with oil and this prevented them from flying. An international bird research centre tried to save some of these beautiful creatures, but the washing treatment itself was traumatic, so they only attempted to wash the strongest birds.

Birds Fleeing from Gulf War

NICOSIA, Cyprus – The Gulf War is not for the birds.

Cyprus bird-watchers say that since the air war began in January, the Mediterranean island has become a haven for birds normally residing in the Persian Gulf.

'The ecolological destruction of the birds' natural habitat in the Gulf Region has forced cranes, red-breasted geese, mule swans and white storks to search for a safe place,' said Pavlos Neophylou, secretary of the Cyprus Ornithological Scoiety.

Neophylou said the Society had observed birds' migratory patterns shifting to Cyprus during Mid-East conflicts as long ago as the beginning of the Iraq–Iran War in 1980 and the civil war in Lebanon.

Associated Press, *The Gazette*, Montreal,
2 March, 1991 K10

The Persian Gulf coastline has sandy beaches, small green areas, mud flats and estuaries – habitats for a large variety of biota. Along the coast there are mangrove trees, which serve as a nursery for both shrimp and fish. Coral reefs support fish hatcheries as well as sustaining the endangered Hawksbill turtle. Other endangered species, like the Siberian crane, the Dugong (similar to the Florida manatee), and green turtles feed on the undersea grasses. Common cranes, geese, herons, pelicans and ducks all migrate through the Gulf, and porpoises are found in the straits between Qatar and Iran. The Gulf is not very deep, only about 39 metres (110 feet), and water exchange is very slow, so the crude oil quickly coated the sea bottom, destroying the breeding ground for much of this marine life.

The oil spills in the Gulf also threatened to clog water desalination plants and electrical generating plants along the coast. One of the oil slicks hit the tip of the island of Abu Ali – home to tens of thousands of migrating birds. The nearby desalination plant processed 870 million litres of fresh water a day. The first desalination plant to close due to crude oil was at Safaniya on the Saudi coast. This slick was probably caused by one of the smaller spills, possibly by US attacks on an Iraqi tanker. The Allies claimed that Iraq had caused the biggest spill, by opening the taps at the Mina al-Ahmadi terminal in occupied Kuwait. Iraq claims that bombing destroyed both its tankers and the pipelines. One wonders if it really matters which side committed these environmental crimes; in the end both were guilty of waging a war that paid little heed to the natural surroundings.

One of the most enduring images of the Gulf War was the enormous clouds of toxic smoke caused by hundreds of burning oil fires. In *Global Environment Change*, a newsletter based in Arlington, Massachusetts, Brad Hurley estimates that as many as 1200 fires were burning, including oil wells, refineries and storage tanks, with about 1.8 million barrels going up in flames each day. Kuwait reported 'oil lakes' several kilometres across and more than a meter deep created by spilled oil. Some of these had to be deliberately torched to prevent build-up of volatile gases. In one case vehicles crossing such a pool of oil ignited the underlying gases and five people were killed.[61]

On 6 March 1991 US scientist Carl Sagan said that the black rain and smog from burning Kuwaiti oil wells was likely to cause massive crop failures throughout the Middle East and south Asia. He theorised that the smoke would block out the sun's rays, thereby reducing temperatures over large areas of the Earth and disrupting the monsoons. These rains are produced when warm summer air on the continent rises and pulls moisture from the oceans. Other scientists disputed his theory, claiming that as the smoke dispersed over the Tibetan plateau it would be much closer to the ground. Therefore the heat it emitted in the lower atmosphere would have a warming effect, which would be greater than the high atmospheric effect of cooling. These scientists predicted an enhancement of the monsoons. According to another environmental model developed at the Max Planck Institute in Germany, the warming might cause the monsoons to arrive earlier and more forcefully than usual.

What was observed in time was a huge typhoon that struck Bangladesh on 1 May, killing more than 100,000 people. Typhoons are not uncommon in Bangladesh, but this one was accompanied by unusually severe flooding – two feet higher than had ever been previously recorded – and it was followed by exceptionally heavy rains. The speculation as to whether or not the violent storm was related to the fires has never been publicly resolved, although there may well be classified papers on this subject.

The British Meteorological Office predicted that the smoking oil fires would produce acid rain for up to 2000 kilometres downwind of Kuwait.[62] Soviet scientists reported very high levels of acid rain in southern Russia. Satellite images showed smoke and darkened snow in Pakistan and northern India. Astronauts aboard the space shuttle Atlantis reported that they had never seen so much haze shrouding the Earth. It was especially thick over central Africa, where they could barely see the ground.

By August, a giant waterspout and fierce storms had lashed out near a Black Sea resort, killing more than 30 people and causing thousands to flee. There were heavy storms, flooding, and landslides in the mountains that backed these seaside towns. Researchers from the Chinese Academy of Sciences claimed that dense clouds from the Gulf War were also responsible for the disastrous floods which took place in their country. According to Zeng Qingcun, director of the Academy's Institute of Atmospheric Physics: 'The abnormal phenomena, worsened by Mount Pinatubo's eruption in the Philippines, has led to non-stop rains in the Yangtze and Hui river valleys.' In Burma, seven towns were submerged and 200,000 people left homeless.

Although Mount Pinatubo may have had some effect on China

and Burma, it likely did not influence the simultaneous record heavy storms over Eastern Europe. The Philippines lie between the equator and 20 degrees north latitude, and anything emitted into the air there is primarily affected by the trade winds that circulate between the equator and the Tropic of Cancer. Eastern Europe lies between 40 and 60 degrees north latitude and is primarily affected by the prevailing westerly winds that blow towards Russia. There was severe flooding from Bavaria to Czechoslovakia, with several deaths, destruction of farmland, and bridges washed away. Rail lines throughout Austria were submerged and the Danube River reached record heights.[63]

Within 50 kilometres of the fires, temperatures dropped by as much as 20 degrees centigrade. The massive worldwide study of the effect of the fires led many scientists to call it the 'worst man-made pollution event in history'. Dr Richard Small, an atmospheric scientist at the Pacific Sierra Research Corporation in California, called it 'a tragedy for the region, but extremely important for science'. The data collected was useful for studies on acid rain, global warming, ozone depletion and other atmospheric phenomena. The World Meteorological Organization (WMO) in Geneva hosted the first scientific conference on the oil fires in April of 1991.[64]

In the popular press, the fires were usually dubbed the 'Kuwait oil fires' and most were believed to have been caused by Iraq. The Iraqi forces did indeed torch oil fields as they retreated from Kuwait but Iran was said to be experiencing 'repeated black rain events' as early as 22 January 1991. Also satellite images made in mid-February by the US Landsat-5 and the US National Oceanic and Atmospheric Administration (NOAA) in Boulder, Colorado, revealed 'smoke plumes several hundred kilometres long emanating from various regions of Iraq'.[65] Laura A. Gundel, an aerosol expert at the Lawrence Berkley Laboratory, noted that the first suspicious 'spikes' of soot were measured at the Hawaiian NOAA observatory, Mauna Loa, in early February. Both of these observations took place long before the Iraqis' withdrawal from Kuwait and suggest that some oil fires were caused by the Allied carpet bombing.[66]

A scientist at the US Department of Energy's Lawrence Livermore Laboratory, was asked not to deliver their computer simulation of the fires at a scientific conference in Vienna. According to William Arkin, a US national security expert, officials at the Pentagon and State Department told him privately that they were concerned that:

> such revelations will spur demands that Saddam Hussein be tried for environmental war crimes. These sources suspect that Hussein would defend himself under 'military necessity' exclusions: smoke from burning oil wells, for example, concealed Iraqi troops from Allied bombers. The officials worry that this defense might then lead to calls for stricter environmental protection provisions in the international rules of war.[67]

On 30 January 1991, just as the Gulf War was building up momentum, the White House waived the legal requirements for assessments of the effect 'Pentagon projects' might have on the environment. This was done because of 'concern that war efforts could otherwise be hampered'. The Pentagon assured the public that 'the military had no intention of misusing and enlarging the exemption to accomplish other objectives than to test new weapons, increase production of materials, and start new activities at its military bases'.[68]

An international treaty protecting natural ecosystems, the United Nations Convention on Environmental Modification, had been signed in 1977 after the Vietnam War. In this conflict, the military had deliberately targeted the environment as a strategy of war: more than two million acres of forest were levelled and denuded; another five million acres of land were left contaminated and unproductive because of toxic chemicals such as 'Agent Orange'. It is estimated that land sprayed with this defoliant cannot be used for at least a hundred years (see p157). During the war in Burma, the military launched a scorched-earth policy to sweep guerrillas from the beautiful teak forests. This was also done in South Africa where, in the process, herds of elephants were destroyed. The new treaty banned all such deliberate attempts to wreak ecological havoc: blowing up dams, bombing chemical or nuclear plants; releasing chemical contaminants into the air; causing earthquakes or tidal waves; burning forests or crops; or contaminating water supplies. It is said that the convention was initiated because of the outrage caused by US attempts to seed clouds in Vietnam to disrupt the weather patterns.

However, according to the international lawyer Richard Falk at Princeton University, the convention does not cover the environmental holocaust resulting from a nuclear war because the intention would be to destroy enemy capabilities, and the side effects, namely contamination of the air, water and land, are not intended. By the same token, dropping bombs on Iraq and causing oil spills and fires is also not covered. It is clear that the 1977 treaty is inadequate for today's warfare.

The War Goes On

President George Bush officially declared a ceasefire on 28 February 1991. Yet, even ten years later, NATO forces are still conducting bombing raids. In June 2000, the *Guardian* reported that this 'low intensity warfare' was escalating: an estimated 78 tonnes of weapons had been dropped on Iraq by British aircraft since December 1998, compared to 2.5 tonnes over the six previous years. The Liberal Democrat foreign affairs spokesman, Menzies Campbell, commented:

> There is now persuasive evidence that there is an attritional campaign against Iraqi ground-based air defence systems that has gone beyond the purpose of the no-fly zones ... This represents a significant policy shift, which has never been announced or explained to parliament.[69]

International sanctions against Iraq have not been fully lifted. Despite a UN programme in which Iraq can exchange oil for food and humanitarian aid, ordinary civilians are still suffering because the West remains highly suspicious of Saddam Hussein's intentions. The lifting of sanctions was further delayed by a botched UNSCOM (United Nations Special Commission) mission sent into Iraq to destroy all 'weapons of mass destruction'. While they did accomplish some important positive goals, their effectiveness suffered when they were accused of harbouring spies. Apparently both British intelligence and the US CIA gained access to UNSCOM without the knowledge of its director, Richard Butler, planting listening devices in sensitive Iraqi buildings and gathering military intelligence.[70] This meant that Saddam Hussein was reluctant to go ahead with weapons inspections.

At a packed meeting in Kensington, London, in early 2000, journalist John Pilger stated: 'According to Unicef, Iraq in 1990 had one of the healthiest and best-educated populations in the world; its child mortality rate was one of the lowest. Today, it is among the highest on earth.' It was Pilger's disturbing film, *Paying the Price: Killing the Children of Iraq*, broadcast on British television in March 2000, which has done more than anything else to galvanise the public. The Foreign Office has reportedly been shaken by the massive outcry against the sanctions.[71]

Allied veterans of the war are still being affected by 'Gulf War syndrome', and their campaigns to gain recognition and compensation continue. Many believe there is a link between the symptoms they experience and the use of DU ordnance. In a letter to President Clinton dated 8 June 2000, Tony Hall, a member of US Congress, joined with other congressional leaders in calling for an investigation into depleted uranium's effects on human health:

> Veterans of the Gulf War are fighting health problems which are not well understood by medical professionals but which are real and affect their lives in significant ways. There are credible reports that Iraqi civilians are suffering too: cancer rates appear to be significantly higher than the worldwide average … The suspected culprit is depleted uranium, a toxic and radioactive metal. Yet, nine years after the Gulf War ended, few efforts have been made to examine its effects on human health.[72]

On a very human level, returning soldiers were left to deal with the trauma of what they had seen and done. One soldier, whose company had been involved in one of the 'friendly fire' incidents, described how his life simply fell apart:

> Everyone was eaten up with guilt and about 30 per cent of my company were sick. There was no leadership or counselling, we were all fighting with each other, drinking too much and getting

sicker as time passed. My wife was sick and hurting so bad she was totally homebound. I was discharged out of the Army on 31 December 1991. My records, medical and personal, were lost. I had become a model soldier before and during the war, but after the war was a different matter.[73]

United States National Gulf War Resource Center Executive Director Responds to a Letter on Gulf War Syndrome in the Wall Street Journal*:*

Dear Editor,

There are 183,629 Gulf War veterans who have filed claims against the Department of Veterans Affairs for service-related disabilities. Of those, 136,031 were approved. More than 263,000 War veterans have sought care at the Veterans Administration. Some 9,600 have died.

These figures are from the 576,000 Gulf War conflict veterans eligible for healthcare (those serving from August 2, 1990 until July 31, 1991).

In addition, according to the Department of Defense, approximately: 100,000 US troops were exposed to low levels of chemical warfare agents, including sarin, cyclosarin, and mustard gases; 250,000 received the investigational new drug pyridostigmine bromide; 8,000 received the investigational new botulinum toxoid vaccine; 150,000 received the hotly debated anthrax vaccine; 436,000 entered into areas contaminated by 315 tons of depleted uranium radioactive toxic waste possibly laced with trace amounts of plutonium; and hundreds of thousands experienced the hell on Earth of more than 700 burning oil well fires in and near a combat zone for months.

According to the *Washington Post* today, some 1,200,000 Iraqi civilians have died since the start of the Gulf War—a war with no end in sight for civilians or soldiers.

Sincerely,
Paul Sullivan
Executive Director

REMEMBERING THE PAST; QUESTIONING THE FUTURE

We have analysed two wars, both of which were officially justified as 'humanitarian' causes. On reflection, can we really call the suffering inflicted on the civilian populations of Kosovo and Iraq humane? Over the centuries, civilised societies have developed and promulgated prisoners' rights and anti-torture laws, but there are no limits as to the mistreatment of nations that fall from international grace. This designation of 'rogue nations' by the international community is a new phenomenon, but there are some rules in international law that could apply:

1. Starvation of civilians as a method of warfare is prohibited.
2. It is prohibited to attack, destroy, remove, or render useless objects indispensable to the agricultural areas for the production of food stuffs, crops, livestock, drinking water installations and supplies, and irrigation works, for the specific purpose of denying them for the sustenance value to the civilian population or to the adverse Party, whatever the motive, whether in order to starve out civilians, to cause them to move away, or for any other motive.[74]

Russia is struggling with its own complex problems of war and governance. The International Assembly for Human Rights Protection, a Russian NGO, convened a meeting on 15 May 1996 to deal with many of the same questions raised here. The conference called for a broad analysis of the causes of the Chechen War and the suffering that resulted. They made 16 concrete proposals for national healing. I was especially touched by some of them:

- to take public control over the definition of damage caused by military operations and over compensation to those who dwell in the combat zone;
- to find possible ways for psychological rehabilitation for all people involved in the conflict;
- to establish local resource and human rights protection centres, and to prepare a Book of Memory for those who were killed;

- to suggest to the President of the Russian Federation that he make more use of Russian NGOs in settling crises in and around the Chechen Republic.[75]

Dealing with the after-effects of war is a positive process, but these proposals also emphasise the need to look at the problems that give rise to violence. I would liken war to the eruption of 'tumours' that tell us of pre-war malignant roots. The Kosovo crisis and the Gulf War were as much results of economic manoeuvring, political expediency and failed diplomacy as they were of more apparent humanitarian concerns.

In terms of the weaponry used, wars take root in the research labs and military testing sites. Experimentation has an immediate effect that can be observed, and this can help us think about what it would be like to use such a weapon in war. Domacio Lopez of the US Rural Alliance for Military Accountability, for example, has suffered from living downwind of a US depleted-uranium testing range since the 1970s. Based on his experience, he estimates that during the first eight months of the Gulf War, 50,000 Iraqi children would have died from various diseases related to the use of depleted uranium.[76]

It is clear that military research gives us clues as to the nature and outcome of future conflict. Therefore in the next section we will take a close look at the type of research that is currently being carried out and the observable effects of past military experimentation. We will also try to gauge the cost to humanity of diverting money and natural resources into war making. I hope that a strong case will be made for treating military activity as a 'cancer on the body politic' and that we will realise that we have viable alternatives.

PART II
RESEARCH

CHAPTER 2
SEARCHING THE SKIES

In thinking about Earth as a complex living organism, taking in nourishment from the sun,[1] we note that it actively maintains a stasis, or balance, in the composition of the atmosphere and the salinity of ocean water. It maintains its temperature within narrow and predictable bounds, and it sustains an incredible range of plants and animals.

The biosphere, the layer of Earth capable of supporting life, consists of air, water and soil and is about 16 kilometres wide, extending into the atmosphere, underground and underwater. It has survived for millennia by recycling nutrients; through complex interactions of organisms, some utilising the waste of others, the biosphere has formed a functional, sustainable and interdependent whole. However, the biosphere fails to have an efficient excretion system for the complex toxic products that humans have freely created and dispersed into it. Introducing unnatural materials into Earth's system can be compared to introducing 'non-food' into the human body – the body will struggle to rid itself of the unwanted invasion.

In many ways it is fortunate that the biosphere does not always incorporate man-made materials into its system – there are chemical compounds, isotopes and physical states of matter that would have an extremely adverse effect on life if they became part of our interdependent cycle. But it is not only the introduction of toxic materials and other waste that endangers the balance of the Earth's self-regulating system. Recent military research and experimentation has gone still further by manipulating the layers of Earth's atmosphere which separate the biosphere from solar and cosmic debris, protecting it from harmful radiation. Much of this research plans to use planet Earth itself as a weapon, harnessing the power of natural processes for war. To me, this is one of the most disturbing, least understood military abuses of the environment.

In order to understand the natural processes military research is exploiting, we need a brief explanation of the layers surrounding the Earth. In focusing on how these layers function and interact, we can see the dangers of interrupting the natural balance and enhance our ability to restore the Earth to health.

THE LAYERS OF THE ATMOSPHERE ABOVE EARTH

In trying to describe Earth and its protective layers, one has to imagine the planet rushing through space at an incredible speed, about 107,280 kilometres (66,600 miles) per hour, on an elliptical path around the sun. In addition to moving at this incredible speed, Earth is turning on its axis once every day. All this movement means that the layers of the atmosphere do not stay at the same distance all over the globe at all times: the forward side of the Earth will have a slimmer atmosphere and will trail behind to the rear; atmospheric layers are generally closer together at the magnetic poles and further apart at the equator; the atmosphere is also impacted by changes in the sun, moon and cosmos. In the following descriptions, the measurements I have given are approximate for the northern temperate zone.

The troposphere: As altitude above the Earth increases, the temperature drops until a minimum is reached. Then it begins to get warmer again higher up. The layer of Earth's atmosphere from the surface to the first minimum temperature level is called the troposphere or lower atmosphere, and the minimum itself is called the tropopause (approximately 10 kilometres above the Earth, or 33,000 feet). Commercial aeroplanes today fly at around this height.

In 1993, Reginald E. Newell of Massachusetts Institute of Technology reported the discovery of large rivers of water vapour in the Earth's troposphere that rival the Amazon in their size and flow rate. There are five such rivers in the northern hemisphere, and five in the southern hemisphere. They are 676 to 773 km (420-480 miles) wide, up to 7700 km (4800 miles) long, and flow in a narrow

The Layers of the Earth's Atmosphere

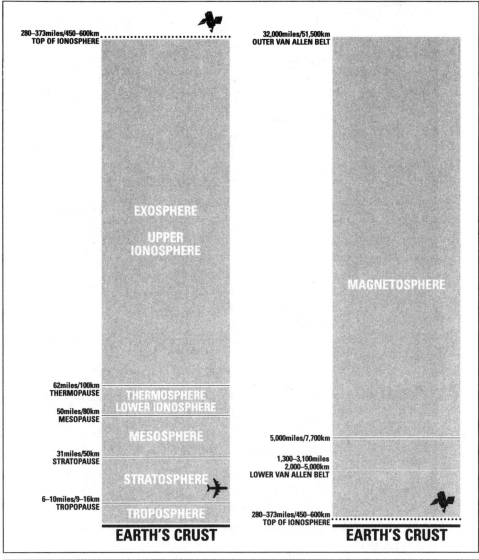

280–373miles/450–600km
TOP OF IONOSPHERE

32,000miles/51,500km
OUTER VAN ALLEN BELT

EXOSPHERE

UPPER IONOSPHERE

MAGNETOSPHERE

62miles/100km
THERMOPAUSE

THERMOSPHERE
LOWER IONOSPHERE

50miles/80km
MESOPAUSE

MESOSPHERE

5,000miles/7,700km

31miles/50km
STRATOPAUSE

1,300–3,100miles
2,000–5,000km
LOWER VAN ALLEN BELT

STRATOSPHERE

6–10miles/9–16km
TROPOPAUSE

TROPOSPHERE

280–373miles/450–600km
TOP OF IONOSPHERE

EARTH'S CRUST

EARTH'S CRUST

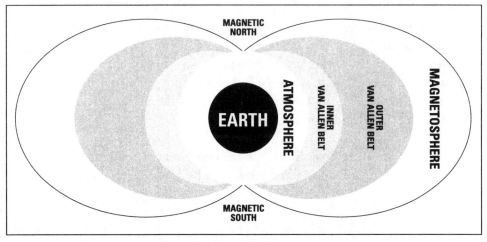

MAGNETIC NORTH

EARTH

ATMOSPHERE

INNER VAN ALLEN BELT

OUTER VAN ALLEN BELT

MAGNETOSPHERE

MAGNETIC SOUTH

band only 3 km (1.9 miles) above the Earth's surface. They are the main vehicles for moving water around the globe – for example, from the tropical rain forests at the equator to the temperate zones – and as such, have a major influence on climate, weather patterns and water distribution. There is speculation that interfering with these rivers could induce floods and droughts.

The stratosphere: Above the tropopause is the stratosphere, which continues to approximately 50 km (31 miles) above the surface of the Earth and, surprisingly, keeps getting warmer as distance increases. One of the reasons for this increase may be that at approximately 25 km (16 miles), the stratosphere contains the ozone layer. Ozone molecules contain three oxygen atoms, instead of the two atoms in normal air. At Earth's surface we consider this ozone to be a pollutant, one of the major components of smog, but in the stratosphere ozone captures the sun's ultraviolet rays that can cause damage to crops, animals and humans. Military planes usually fly in the stratosphere at around 50,000 feet.

The jet stream: During World War II bomber pilots found that they experienced a great deal of turbulence when flying in the lower stratosphere (commercial planes flew at much lower heights at this time). It was discovered that there were narrow bands of high-speed winds encircling the Earth at altitudes of around 10 km (33,000 feet) or above. These were named the 'jet streams'. With the discovery of the jet streams, humans began to understand that the Earth's atmosphere and weather were more complex than had ever been imagined.

The mesosphere: As the temperature increases with increased altitude in the stratosphere, it reaches its highest point, called the stratopause. Above this is the mesosphere, in which the temperature again decreases with increased height above the Earth. The mesophere reaches its lowest temperature at the mesopause, about 80 km (50 miles) above the Earth.

The ionosphere: Above this limit are two layers of the atmosphere that are normally considered together, the ionosphere. The lower part of the ionosphere is called the thermosphere, and it stretches between 80 and 100 km (50–62 miles) above the surface of the Earth. This layer, as its name implies, is very hot, reaching temperatures of between 300 and 1700 degrees centigrade (572–3000 degrees Fahrenheit). Away from Earth the temperature of the thermosphere rises to its highest point. The upper part of the ionosphere is called the exosphere, and it extends from about 100 to 600 km above the surface of the Earth; its temperature cools with distance from the Earth. It is in this outermost layer that most satellites go into orbit.[2]

While most of the lower atmosphere and Earth itself is electrically neutral, the ionosphere is electrically charged and capable of carrying an electric current. The German mathematician Carl Friedrich Gauss speculated that there might be such a region as early as 1839. In 1902, the American engineer Arthur E. Kennelly and the British physicist Oliver Heaviside echoed the idea in order to explain the fact that radio waves could be reflected off a layer of atmosphere and sent around the curved surface of the Earth. This theory was proved in 1925, and for a while the lower ionosphere was called the Kennelly-Heaviside layer of the atmosphere.

Modern astrophysics tends to classify the ionosphere as a whole, with three distinct regions based on the degree of ionisation rather than temperature. The first or lowest is the D layer – it is strongly ionised during the day but not at night. It is used primarily for short wave and broad-band radio transmission. Above this is the E layer, extending from 90 to 140 km (56–90 miles) above Earth's surface. This layer has ionised molecules and strong electrical currents. Above 140 km, the F region, sometimes divided into F1 and F2, contains ionised atoms, with the F2 portion containing the highest concentration of ions. The E and F regions are responsible for long-distance propagation of radio signals.

Ionisation occurs when solar rays cause the electrically neutral atoms that exist at this high level to lose electrons. The electron then has a negative charge and the atom a positive charge. These charged particles are called positive ions and negative ions. In the natural Earth environment, we do not experience this kind of electrically charged atmosphere, technically called a plasma, except briefly after a thunder and lightning storm. Most of the atoms that surround us are electrically neutral, and matter exists in solid, liquid or gaseous state. Plasma is a superheated gas, sometimes called the fourth state of matter.

The ionosphere is one of the most important protective layers that blanket Earth, shielding us from damaging solar and cosmic particles.

The electrojet: The ionosphere contains two very large rivers of direct current electricity, called the electrojet, which circulate within the ionosphere at heights of about 120 km (60 miles), dipping lower at the north and south poles. Like the jet stream and the massive rivers of water vapour, the electrojet moves particles around the whole planet. It is a source of electrical power that surpasses anything on Earth.

The magnetosphere: Beyond the ionosphere there is another discernible layer of protective Earth cover, called the magnetosphere. In this region, the Earth's magnetic field largely controls the energetic movements of ions; above this, in 'outer space', particles are controlled by the sun's field. The magnetosphere contains giant magnetic force lines running between the magnetic poles, called the Van Allen belts.

The Van Allen radiation belts: The Van Allen belts were named after an American physicist, James Van Allen, who discovered them in 1958. The lower Van Allen belt occurs between 2000 and 5000 km (1300 and 3100 miles) above the surface of the Earth, sometimes referred to as within one Earth radius (the average radius of Earth is about 6500 km or 4000 miles). This rough measurement helps the imagination. The outermost structure protecting our Earth, the outer Van Allen belt, is about 51,500 km above the surface of the Earth or within 8 to 9 Earth radii. Because of the spin and wobble of the Earth, the innermost Van Allen belt dips down to within 200 km (120 miles) of Earth above the southern Atlantic ocean (known as the South Atlantic anomaly). There is a similar anomaly above Mongolia.

The stream of charged high-energy particles emitted by our sun is called solar wind. The solar wind is strongest during sun flares and sun spot activity, sometimes disrupting radio transmission on Earth and causing a great deal of static. Solar and cosmic particles are hurled towards the Earth and are captured in the Van Allen belts, spiralling around the magnetic force lines that stretch between the north and south poles in as little as 0.1 to 3 seconds. The trapped particles tend to stay in the magnetic field for a very long time, unless the fields are seriously disturbed – for example, during magnetic storms. Sometimes charged particles avoid capture by the Van Allen magnetic lines and reach the Earth's upper atmosphere at the poles. As the energetic particles strike the gases of the atmosphere they 'glow', causing the beautiful displays of coloured lights known as the Aurora Borealis in the northern hemisphere and the Aurora Australis in the south.[3]

The moon: The moon is held in orbit by Earth's gravity at approximately 384,000 km (240,000 miles) from the Earth. It has no atmosphere and experiences extremes of temperature – about 100 degrees centigrade when fully exposed to the sun, and about -200 degrees centigrade (212 to -328 degrees Fahrenheit) during the lunar night.[4] Earth's gravity effectively extends out a distance of 1.6 million km; beyond this an object would drift into an orbit around the sun.

The sun: The sun is 148 million km (93 million miles) from Earth, yet it supplies a constant source of energy that keeps Earth 'alive'. Seasonal variations occur because of the orbital rotation of Earth and the tilt of its axis. All of this solar energy is reflected back to outer space over time as long-wave radiation, maintaining a global equilibrium; the temperatures we experience are the balance between incoming and outgoing heat. If this did not occur, Earth would keep getting hotter, destroying all life. Gases, clouds and suspended particles in the atmosphere immediately reflect back about 26 per cent of solar energy, and the surface of the Earth reflects another 4 per cent. Therefore about 70 per cent is absorbed by the Earth and its atmosphere and is redistributed as heat and moisture by the oceans, large bodies of inland water, the prevailing winds, and the trade winds.

Most human space exploration has taken place well within Earth's atmosphere, but there have also been some trips to our moon and outer space. Humans have always been curious about this 'air' above us, our solar system and the cosmos which envelops it, but it was not until the second half of the twentieth century that it was actually possible to visit space and learn about it first hand. This exploration required the development of rocket technology – a technology that had always been linked to war. It is not surprising, then, that the exploration of space also took place under the auspices of the military.

When the space programme was born, the military promised new medicines produced in the weightless environment, new research into the history and evolution of the Earth, even the commercial exploitation of mineral resources on the moon. Space research seemed quite wonderful – creative, inventive, exciting and enterprising. To the casual observer, there was little to indicate that the growth in both knowledge and accomplishment would become malignant.

ROCKETS

The early history of atmospheric research includes the use of a weather vane, the Greek Towers of Wind, and Aristotle's *Meteorologica*, written about 340 BC. It was not until about 1600 AD that Galileo invented the thermometer and 1643 that Torricelli developed a barometer. When it was discovered that weather changes and barometric changes were related, many stations for monitoring barometric pressure were set up. Using these measurements, Benjamin Franklin discovered, around 1743, that storms were travelling atmospheric systems related to high- and low-pressure areas. Franklin could only observe and had not yet learned to predict, much less change or control the weather. During the later half of the nineteenth century balloons were used to probe the upper troposphere, and in 1899, the French meteorologist Teisserence de Bort discovered the stratosphere. With the development of rocket technology, atmospheric research took a giant leap forward.

Rockets were originally called 'Chinese arrows' after the earliest description of their use by the people of Kaieng-fu in 1232 AD to stop an invasion by the Mongols. Tubes that contained an early form of gunpowder were tied to the shaft of an arrow, producing a fiery exhaust. Simple rockets are all based on more or less the same principle. The rocket has a head, usually cone-shaped, which sits above a bursting charge. Below the charge there is a cone-shaped choke, with the smaller opening attached to the charge. The gas exhaust from the exploding charge escapes rapidly down a chute, moving the rocket in the opposite direction.

As understanding of the lower atmosphere improved, so did the technology of rockets. Primitive rockets were 'improved' upon by an English Franciscan monk, Roger Bacon, and by the Germans, Italians and Syrians, but their only function was in war.[5] In the eighteenth century, India tried to repulse British troops with rockets that could travel 2.5 km (1.6 miles). They were constructed with thick stalks of bamboo, 2.5 to 3 metres long, attached to iron tubes.

Use of rockets for space travel was first proposed in 1883 by a Russian schoolteacher, Konstantin Tsiolkovsky. Commercial jet engines require oxygen, which they take in and compress, inject with fuel and ignite to form combustion. Jet planes cannot travel at heights where the oxygen density is too low. So, in order to travel into space, humans needed a form of jet propulsion that carried its own oxygen. Tsiolkovsky posed the idea that an internal combustion engine containing liquid oxygen with either liquid hydrogen or liquid kerosene could be used to replace the need for atmospheric oxygen. He imagined that his space rockets could carry passengers and thought that they would need double walls to protect them from meteoroids. It was his idea to have a series of step rockets that could be fired and then discarded, each step boosting the rocket farther aloft. He also thought of using gyroscopes for stabilisation, another feature found in today's rockets.

Actual construction of liquid-fuelled rockets was pioneered by the engineer Robert H. Goddard, after whom the NASA Space Center outside Washington DC was named. In 1926 he successfully launched the world's first such rocket and later, in Roswell, New Mexico, he built a rocket that could reach 90 metres (300 feet) above ground. At the same time, Germany was developing its rocket technology. The Treaty of Versailles (1919) had forbidden Germany from building aircraft, but there was no explicit prohibition against rockets. Under the leadership of Werner Von Braun, a young rocket enthusiast, the German army agreed to fund trials of experimental liquid-fuelled rockets. After some unsuccessful tries, Von Braun and his colleagues finally began to achieve some flights.

The original site of Von Braun's rocket testing was later returned to the military, to become an ammunition depot, and a major rocket research facility was built near the village of Peenemunde, on the Baltic coast. The large A-4 rockets used to bombard London, Antwerp and other cities in 1944–45 were built here. The facility also built the V-1 and V-2 (Vengeance 1 and 2) – small pilotless planes that delivered death and destruction to London and southern England. The V-1 had a range of 220 km and the V-2 a range of 3200 km. After the Second World War, Von Braun and others surrendered to the Americans and offered to carry on developing their rockets in the United States.

From 1945 to 1955, rocket technology continued to be defined by the military. Rockets were refined for use as antitank weapons, intercontinental ballistic missiles, aircraft interceptors and for high-altitude research.

The First Space Rockets and Satellites

The first artificial satellite was launched into space by the Soviet Union on 4 October 1957. It was called Sputnik 1, which can be translated as 'Fellow Traveller 1'. When Sputnik 2 was launched on 3 November 1957, it contained a dog – Laika. Laika survived for several days but died of heat exhaustion before her oxygen supply ran out. The dog captured the imagination of the world, and dreams of human space adventure were born.

The US hurried to place its Vanguard 1 satellite in orbit, in December 1957, but it exploded shortly after lift-off. It was Explorer 2, launched on 31 January 1958, which was the first successful US space venture and which was credited with the discovery of the lower Van Allen belt. Surprisingly, the Geiger counters on board both the US and Soviet satellites had failed to measure radiation at about 1000 km (620 miles) above the Earth and beyond – the radiation at this height was so intense it was off the scale. It took some time before each country's scientists, working separately because of the Cold War, realised this. The harmful radioactive particles coming from the sun and cosmos were discovered, trapped high in the Earth's atmosphere.

On 3 December 1958, the US satellite Pioneer 1 reached a height of 110,000 km (70,000 miles) and discovered the outer Van Allen belt. A Soviet researcher, S.N. Vernov, discovered the same phenomenon based on data sent back from Sputnik 2.

The most obvious target of all of this very public Cold War activity was the moon. The Soviets tried to hit it, but missed with their Luna 1 – however, it was the first rocket to escape Earth orbit. On 14 September, Luna 2 crashed on the moon, and in October 1959, Luna 3 managed to orbit the moon, sending pictures back to Earth. Some of the interest in the moon is in its potential as a 'resting station' on the way to other planets in our solar system. Perhaps it was also envisioned as a military base, although it cannot be claimed by any one country. Talk of mining minerals on the moon is still current even though the United Nations Treaty on Outer Space forbids depleting non-renewable lunar resources.

NO LONGER JUST OBSERVERS

Atmospheric exploration moved rapidly from pure observation to experimentation. The US started nuclear atmospheric testing in the Pacific in 1946 and in Nevada in 1951. By the end of 1956 the US had set off more than 86 nuclear bombs. The Soviets began atmospheric testing in 1949 in their Arctic region, and by the end of 1956 had set off 15 explosions. The UK detonated a further nine atmospheric blasts on the Monte Bello Islands off western Australia and near Maralinga in southern Australia.[6] Experiments on the ionosphere began almost immediately after the Van Allen belts were discovered – before we even knew the role they played in protecting the Earth.

Project Argus (1958)

Between August and September 1958, the US navy exploded three fission-type nuclear bombs 480 km (300 miles) above the south Atlantic Ocean. As noted earlier, the Van Allen belts dip to 200–400 km at this point, often disrupting ship communication. In addition, two hydrogen bombs were simultaneously detonated at 160 km over Johnston Island in the Pacific. Johnston Island is about halfway between Hawaii and the Marshall Islands, 18 degrees north of the equator, yet these blasts were so high in the atmosphere that they could be seen in Tahiti, French Polynesia, about 18 degrees south of the equator. The experiment, under the code name Project Argus, was designed by the US Atomic Energy Commission and the US Department of Defense, who called it 'the biggest scientific experiment ever undertaken'.[7]

The purpose of Project Argus appears to have been to assess the impact of high-altitude nuclear explosions on radio transmission and radar operations. Through previous atmospheric explosions, the military had discovered that nuclear bombs create an electromagnetic pulse (EMP) that wipes out radio communication. The navy also wanted to increase understanding of the ionosphere and the behaviour of charged particles within it. It is possible that the newly discovered layers of the atmosphere were seen as potential sources of unlimited energy and destructive power.

These nuclear explosions created new magnetic radiation belts and injected sufficient electrons and other energetic particles into the ionosphere to cause worldwide effects. It is not known how long these belts lasted but they were observed five years after the explosions. The electrons travelled back and forth along the newly created magnetic force lines, causing artificial 'auroras' when striking the atmosphere near the north pole. This phenomenon was apparently what inspired the concept of a space 'shield' against incoming missiles. If natural Earth processes destroyed incoming debris, could an artificial shield be created against intercontinental missiles?

The effects of Project Argus on Earth have never been fully revealed. However, by the winter of 1957 it was clear that nuclear testing was causing severe problems for people living near the magnetic north pole.

The Year the Caribou Did Not Come

The Inuit people of Baker Lake, near Hudson Bay in Canada's Northwest Territory, explained to me that they were 'newcomers' and had only lived in a village for the last 30 years. That was in 1988. During the winter of 1957–58, the caribou, upon which the Inuit depended for food, clothing and shelter, had failed to migrate across the northern tundra – something that had apparently never happened before in the 3000-year oral history of their people. One of the women elders, whose leathery skin spoke of many years of hardship, said something to me in Inuktituk. My interpreter looked at me: 'she says the death came from the sky'. The Inuit People had seen the unnatural aurorae borealis, and some had made a connection between this and the fact that the caribou had failed them for the first time.

Qiayug, one of the survivors, who belonged to the Ahiamuit tribe – literally the 'people who live apart' – told Canada's *Globe and Mail* that although the Inuit had sometimes suffered deprivation over the centuries, 'that winter was the very worst hunger they could remember'. Many starved to death. According to the *Globe and Mail:* 'It remains unknown whether the shortage was due to over-hunting or a scarcity of plant life on the tundra.'[8] The article made no mention of the 'death' from the sky.

In response to the crisis, the Canadian government sent helicopters over the tundra in the spring of 1958, gathering up those who had survived. They established settlements for the Inuit and told them they could not return to live 'on the land'. The settlements consisted of prefabricated houses, foreign to the people and considered by them to be much colder than their icehouses, which had warm fires and floors of caribou skin. Many Inuit now build icehouses around their front doors to protect them from wind, snow and cold – even after all these years, they still feel alienated in these artificial villages.

In 1956, C.E. Miller and L.D. Marinelli of the Argonne National Laboratory reported that the fall-out product caesium 137 was being found in human bodies.[9] Caesium had become incorporated into grasses, vegetables, milk and meat. The problem was especially acute in the Arctic where the caesium tended to concentrate in lichen, the food of caribou over the long Arctic winters. In 1961, Linden Kurt reported that caesium levels in Swedish reindeer were 280 times higher than in beef cattle.[10] The level of caesium in the bodies of Sweden's Arctic people was 38 times greater than in the bodies of southern Swedes. Canadian researchers also noted that caesium levels in both caribou and humans were consistently higher between January and June than between July and December. Caribou calves are born in the spring.

Contamination levels in the region around Baker Lake were the highest measured in a study of the Canadian north carried out by Health Canada.[11] Most of the Inuit and Dene people of this region did not speak English, and even if they did, they were unlikely to be reading the professional science and medical journals. They were not warned about the contamination of the caribou or about their own consistently high levels of caesium 137. The government dismissed the elevated caesium levels as a problem 'which was coming down' as nuclear atmospheric fall-out ceased. Measurements at the height of fall-out exceeded the highly permissive maximum levels recommended by the International Commission on Radiation Protection.[12]

But caesium levels weren't the only evidence that human health had been put at risk. On 30 January 1960, the *Canadian Medical Association Journal* noted that cancer rates in the central Arctic appeared to be up to 20 times higher than in the eastern or western Arctic.[13] The areas of increase were the same as areas known for the greatest intensity of 'Northern Lights' and correlated with the artificial aurorae borealis caused by atomic explosions. Public health officials began to note the extraordinary number of cancers which impaired reproductive capacity, mitigating against the survival of the people. No one seems to have connected this phenomenon with the reproductive problems of the caribou!

Not all of the effects of exposure to radiation are immediate; damage to cells manifests itself in various ways over time.[14] By 1975, 16 years after the fateful year when the caribou failed to appear, cancer rates in the central Arctic had risen from 78.4 per hundred thousand to 169.3 per hundred thousand.[15] This increased rate was not due to extended life span, since an increase occurred in all age groups. There was a noticeable growth in the amount of lung cancer, which some blame on the introduction of cigarettes to the Arctic. However, this cannot explain why rates were not highest in the western Arctic, where cigarettes were first introduced. Nor can it account for the higher lung cancer rates in women, when 20 per cent more men than women were known to be smokers. Soviet reindeer herders also experienced the same nuclear legacy. The cancer rate among natives of the Chukotka Peninsula, in the northeastern portion of Siberia, was reported to be two to three times higher than the national average, and according Dr Vladimir Lupandin, the doctor quoted in the article, almost every household had someone with cancer, 90 per cent of the population had chronic lung disease, and infant mortality was one in ten live births.[16]

More Military Experiments

There was a moratorium on atmospheric nuclear testing in late 1958, but that did not stop further experimentation with the ionosphere. In 1961, *Keesings Historisch Archief* reported that the US military planned to create a 'telecommunications shield' in the ionosphere in order to counteract the interference to radio communication caused by solar wind.[17] The plan was to bring into orbit 350,000 million copper needles, each 2–4 centimetres long. Researchers hoped that these trillions of needles would form a belt 10 km (6 miles) thick and 40 km (25 miles) wide and that the needles would be distributed about 100 metres apart. They could then bounce radio waves off this artificial shield rather than the 'unreliable' ionosphere. Although this project focused on problems with radio communication, it again reflects the idea of a space shield, which was rapidly gaining ground in the minds of military planners.

The military actually did try this experiment, tossing 350,000 million copper needles into orbit![18] Since this experiment is not something the military later bragged about or one they expanded as planned, we can assume it was 'unsuccessful' for their purposes. What damage it did to the complicated upper atmosphere is unknown. One independent researcher, Leigh Richmond Donahue, along with her physicist husband, Walter Richmond, tracked such events in the postwar years. She wrote:

> when the military sent up a band of tiny copper wires into the ionosphere to orbit the planet 'so as to reflect radio waves and make reception clearer' we had the 8.5 Alaskan earthquake and

Chile lost a good deal of its coast. That band of copper wires interfered with the planetary magnetic field.[19]

While we cannot prove this hypothesis right or wrong, it was put forward by serious scientists and was the beginning of attempts to connect human disturbance of our atmosphere with unwanted violent happenings on the surface of the planet. Non-military geophysicists were shut out of these experiments. Only small bits of information on the projects were available to the public, yet even with scant information, the International Union of Astronomers strongly opposed the military plan of seeding the ionosphere with copper needles. No one knows what the actual results of this experiment were, and the military aren't telling. The observed effects should be declassified, so that those who are concerned by such experiments can analyse the findings.

Project Starfish (1962)

After a brief respite, the US lifted the ban on nuclear atmospheric testing in 1962 and on 9 July began a further series of experiments with the ionosphere. According to their description, this series would include a 'one kiloton device, at a height of 60 kilometres and one megaton and one multi-megaton, at several hundred kilometres height'.[20] These tests seriously disturbed the lower Van Allen belt, substantially altering its shape and intensity.

In this experiment the inner Van Allen belt will be practically destroyed for a period of time; particles from the belt will be transported to the atmosphere. It is anticipated that the earth's magnetic field will be disturbed over long distances for several hours, preventing radio communication. The explosion in the inner radiation belt will create an artificial dome of polar light that will be visible from Los Angeles.[21]

This was one of the experiments that called forth the protest of the Queen's astronomer in the UK, Sir Martin Ryle, leading him to become a strong anti-nuclear activist.

When I learned of these tests I recalled a story told to me in 1987 by a former Fijian sailor. I met Togea (not his real name) on a lovely Pacific island, in Vanuatu, some distance away from his native Fiji. He was a handsome man of about 40, with that straight posture characteristic of someone who has done military service. Togea attended a lecture I gave and heard me mention the atmospheric testing done in the Pacific between 1946 and 1963. He was very intent when he asked me to come to his home and meet his wife. Something urgent in his eyes made me say yes.

It was a modestly comfortable house, and his wife was obviously surprised at his bringing home a middle-aged white woman. She went to get some tea, her best dishes and cakes, and then settled down to see what all this was about. Very seriously and carefully, Togea told his story, a story he had kept to himself for 25 years. Togea was in the Fijian navy when he was 17 years old, and in July 1962, he was shipped out for a joint exercise with the British navy at Christmas Island. He pulled out a small booklet, produced in the UK, which he said had been given to each navy man and which talked about the Pacific Islands in general and Christmas Island in particular.[22] Christmas Island was a British colony located very near to the equator in the central Pacific. It is now independent, a part of Kiribati. The booklet described the islands as 'uninhabited and generally useless', hence they were being used to test new military weapons – hydrogen bombs. Togea later realised that this statement about the islands being 'useless' was not true, but as a youth he did not question it.

The nuclear test was a super-secret mission, and Togea did not know what to expect. The night before the blast, when their ship was in position, cases of beer were brought out and the men were told to 'drink up' as they might not make it through the next day. Togea could not sleep that night. He thought about dying, and was afraid.

Early the next morning, the hydrogen bomb (described as a multi-megaton nuclear fusion bomb) was exploded; the sea became angry and tossed the ship around like a matchbox. A huge column of thick red and black smoke rose ominously from the small atoll and within a few minutes an ugly roaring fire filled the sky; the ship was directly beneath the 'mushroom' cloud. Togea said he thought it was the end of the world. He asked me if the force of the blast could stop the Earth from turning, as it seemed to him that it had stopped, at least for a moment, on that day. The men were so awed by the event that they said nothing, even to one another. The fire in the sky lasted for three days. Since checking dates and other sources, I am convinced this was Project Starfish.

On arriving back in Fiji, the sailors were warned not to tell anyone of their experience. Togea said that this was not hard. Who would believe them? How could they possibly explain what had happened? He then showed me an open sore on his leg that he had had since that fateful day. No doctor had been able to heal it or even properly diagnose what it was. For Togea, it was a constant reminder of his unspeakable experience. Later, as he learned more about the effects of nuclear bombs and radiation, he wondered if he would develop cancer or if his children would be born with impairments. He was still afraid to speak about it when he married, both because of the official prohibition and because he was worried that if his wife-to-be knew, she would not go ahead with the wedding. Throughout his life happy events were overshadowed by the fear that his exposure to radiation would cause illness in the people he loved – the birth of his two daughters, and later the news that he was to become a grandparent. His wife, who had never known of this deep anxiety and had never had an explanation for his strange behaviour, looked both moved and confused.

For several hours I let Togea and his wife talk out years of events they had never been able to share. I gave what assurance I could, both that their positive experience up to this point was a good omen and that should their worst fears come to pass, neither cancer nor a child with an impairment is the end of the world. However, they and I knew that to cause damage to another person or to the Earth is a crime against life itself. We allowed nuclear bombs to be set off in the sky before we even knew what the sky was and what it did to protect Earth's biosphere, and we exposed Earth's people to radiation long before anyone knew just how dangerous that could be.

The description of Project Starfish in *Keesings Historisch Archief* is less emotionally worded, but still shocking:

The ionosphere (according to the understanding at that time), that part of the atmosphere between 65 and 80 km (40–50 miles) and 280–320 km (175–200 miles) height, will be disrupted by mechanical forces caused by the pressure wave following the explosion. At the same time, large quantities of ionizing radiation will be released, further ionizing the gaseous components of the atmosphere at this height. This ionization effect is strengthened by the radiation from the fission products...[23]

On 19 July...NASA announced that as a consequence of the high altitude nuclear test of July 9, a new radiation belt had been formed, stretching from a height of about 400 km to 1600 km (250–1000 miles); it can be seen as a temporary extension of the lower Van Allen belt.[24]

Later in 1962, the USSR undertook similar planetary experiments, creating three new radiation belts between 7000 and 13,000 km (4300 and 8100 miles) above the Earth. The electron fluxes in the lower Van Allen belt have changed markedly since these high-altitude nuclear explosions and have never returned to their former state. According to American scientists, it could take many hundreds of years for the Van Allen belts to restabilise at their normal levels.[25] About ten years later it was also discovered that the 300 megatons of nuclear explosions set off between 1945 and 1963 had depleted the ozone layer by about four per cent.[26] Hindsight is a great thing but only if we learn from it. These experiments clearly show the danger of undertaking experiments before we have the necessary knowledge to understand the consequences.

The nuclear testing of the 1940s through to the 1960s seriously damaged our environment, but it was also during this time that the people of the Earth formed peace movements that stemmed from a desire to heal rather than harm. Intense civilian pressure forced Britain, the US and the former Soviet Union to sign the Partial Nuclear Test Ban Treaty in 1963, although this was not really the end of atmospheric nuclear testing, since France, China, India and Pakistan were not parties to this treaty and testing continued at lower altitudes, and on a lesser scale, for another 25 years. After the signing of the treaty, nuclear testing in the US, UK and Russia went underground for the most part. The treaty did not forbid either the release of radioactivity into the air or the continued use of rockets to explore space.

Saturn V Rocket (1973)

An accident occurred in 1973 that was to change the course of space experimentation. It again reveals the ignorance of scientists who were experimenting with the ionosphere without understanding it. The Saturn space-launch vehicle requires about 3.45 million kilograms (7.6 million pounds) of propellant just for thrust-off. It then consumes about 12,700 kg (28,600 lbs) of propellant per second for about 150 seconds to give the vehicle a second boost so that it can reach the desired height and velocity. Due to a malfunction in the Saturn V rocket that was used to launch Skylab, the second booster burned unusually high in the atmosphere, above 300 km (186 miles).

This mishap took place over the south Atlantic where the Van Allen belts dip towards the Earth. The burn produced 'a large ionospheric hole', reported by M. Mendillo in 1975.[27] The disturbance reduced the total electron content of the atmosphere by more than 60 per cent over an area 1000 km in radius, and the effect lasted for several hours, preventing all radio communications over a large area. The 'hole' was apparently caused by a reaction between the rocket exhaust gases and ionospheric oxygen ions. This was a surprise to scientists who had thought, or optimist-ically assumed, that rocket gases caused no chemical reaction with the ionosphere.

The booster rocket reaction caused an airglow as radioactive particles struck the gases in the Earth's lower atmosphere – just like the airglow produced by nuclear bombs in the upper atmosphere. These artificial airglows are similar to the natural auroras caused by the sun at the magnetic poles, when particles overflow from the magnetosphere into the ionosphere, but they differ in an important way. Surprisingly, natural auroras are weakest when the sun is most active – the opposite of what one might expect. However, in a time of great activity, the sun releases more ultraviolet radiation than usual, 'strengthening' the Earth's magnetosphere and making it better able to handle the influx of particles. The Van Allen belts, therefore, are not overloaded that easily. Unnatural 'loading' of the magnetosphere does not have this tempering effect because of the lack of increased ultraviolet rays.

After observing the unexpected airglow caused by the Saturn rocket, NASA and the US military began to design ways to test this new phenomenon, recreating it through deliberate experimentation with the ionosphere (their 'Experiments to Understand the Aurora Borealis, through Creating Similar Artificial Luminescence'). These experiments, between 1975 and 1981, were spread around the globe and were followed by more tests involving the newly developed Space Shuttle.

Orbit Maneuvering System (OMS)

During the 1980s global rocket launches numbered about 500 to 600 a year, peaking at 1500 in 1989 (before the Gulf War). The Space Shuttle, introduced during this period, is the largest of the solid-fuel rockets, with twin 45-metre booster rockets. All solid-fuel rockets release large amounts of hydrochloric acid in their exhaust, each Shuttle flight injecting about 187 tons of ozone-destroying chlorine and seven tons of nitrogen, also known to deplete ozone, into the atmosphere. This is in addition to the 387 tons of carbon dioxide released in each Shuttle flight. Soviet aerospace engineer Valery Brudakov has calculated that 300 launches of the Space Shuttle alone could eliminate the Earth's ozone layer in its protective capacity.[28]

In 1981, the NASA Spacelab 3 mission of the Space Shuttle made 'a series of passes over a network of five ground based observatories' in order to study what happened to the ionosphere when the Shuttle injected gases into it from the Orbit Maneuvering System (OMS). The researchers discovered that they could 'induce ionospheric holes', and began to experiment with holes made during the day and at night over Millstone, Connecticut, and Arecibo, Puerto Rico. This artificially induced plasma depletion was then used to investigate other space phenomena, such as the growth of plasma instabilities and the modification of radio propagation paths. The

47-second OMS burn of 29 July 1985 produced the largest and most long-lived ionospheric hole to date, dumping some 830 kg of exhaust into the ionosphere at sunset. A six-second, 68-kg OMS release above Connecticut in August 1985 produced an airglow that covered over 400,000 square kilometres.

It was in 1986 that civilian scientists established the existence of a second ozone hole in the Antarctic. The first ozone hole caused by nuclear testing had begun to 'heal' by then. Why should we worry about this thin band of ozone 40,000 km (25 miles) above our heads? Scientists have estimated that a 1 per cent loss of ozone would result in 1–3 per cent more ultraviolet radiation reaching the Earth. This would, in turn, increase the skin cancer rate and could affect all life forms. It would also change the temperature distribution in the stratosphere, with potential global climate effects.[29] With only a 20 per cent reduction of ozone, humans would experience blistered skin with potential skin cancer development and a depressed immune system allowing the development of other forms of cancer. It is predicted that the incidence of cataracts would increase, food crops would shrivel and burn, shrimp and plankton on the ocean surface would be killed or impaired, and the entire food web of the Earth would begin to collapse. Experts estimate that humans could survive on this impaired Earth for only two years.

We have already noted that the ozone layer in the northern hemisphere was reduced by about four per cent by atmospheric nuclear bomb testing from 1940 to the 1970s.[30] Between 1978 and 1990, the ozone layer in the northern hemisphere decreased by a further 4–8 per cent and in the southern hemisphere by 6–10 per cent.

The effects of rocket gases and atmospheric testing are not the only dangers posed to the environment by the space programme. Nuclear-powered rockets began to thrive in the early 1990s, under then President Bush, who saw them as powerful enough to carry weapons into space, power them for use, and also speed up travel for interplanetary cruisers.

NUCLEAR-POWERED ROCKETS

Projects that the Pentagon wants to keep secret from the US Senate Oversight Committee are called 'black projects'. A limited amount of money is assigned to black projects every year without scrutiny, in the interest of national security. One such 'black project', known as Timberwind, seems to be among the best funded and most detailed of the various rocket schemes that have been leaked to the press. Timberwind is a classified programme involving a nuclear-fuelled rocket, developed at Sandia Laboratory in New Mexico and tested at Saddle Mountain in Nevada. Its plans called for a 75-second suborbital test of a nuclear-powered rocket over Antarctica, and perhaps New Zealand, in April of 1991.

A nuclear rocket does not generate its power by chemical combustion, as do most rocket engines. It heats a propellant, such as hydrogen, in a Radioisotope Thermal Generator (called an RTG), and then expels it at high velocity, providing the forward thrust. It can attain about twice the efficiency of a chemical propellant, powering the rocket at about twice the velocity of other rockets of comparable size. RTGs are fuelled with about 10.9 kilograms (24 pounds) of plutonium dioxide (a ceramic form that is primarily composed of the plutonium-238 isotope), and have to be operational from the moment of launch, meaning that they cannot be launched 'cold' in a safer configuration. A take-off accident would disperse plutonium over a very large area.

Such an accident is a real possibility; the history of the space programme is littered with disasters. The first major space accident to seriously affect Earth occurred on 21 April 1964, when the US rocket SNAP-9A was aborted and the 17,000 curies of plutonium it was carrying were dispersed over a large area of the globe. The plutonium is still detectable in soil and the bones of people and animals. In 1997 there were two SNAP-9A rockets in orbit, each carrying 17,000 curies of plutonium and each planning on completion of their mission to disperse the plutonium as the 1964 rocket had done. There are also the human costs: on 27 January 1967 Apollo 1 burst into flames on its launch pad and all three American astronauts inside were killed; on 24 April 1967, the steering system and parachutes failed on a descending Soyez space capsule and the Russian cosmonaut was killed. Three Russian cosmonauts were killed on 30 June 1971 when a pressure valve mistakenly opened in a descending space capsule. The most shocking rocket disaster happened on 28 January 1986, when the space shuttle Challenger exploded a few seconds after lift-off, killing all seven crew members.

The design effort for nuclear rockets came from the Brookhaven Laboratory in Long Island and Babcock and Wilcox, a private corporation which designed the failed Three Mile Island reactor. The Sandia Laboratory estimated that there was a 4.3-in-10,000 probability that Project Timberwind would crash into New Zealand, releasing large amounts of plutonium. Prior to take-off, the authorities had estimated the likelihood of a Challenger disaster at one in a million.

The Galileo Project (1989)

Another attempt to launch nuclear power into space was the Galileo Project, which carried two RTGs. The spacecraft was built in Germany and was launched on 18 October 1989 from the space shuttle Atlantis. It orbited dangerously close to the Earth twice, but fortunately did not explode or crash, although there was a calculated probability that it might. For unknown reasons, its main antenna would not unfurl and an onboard tape recorder seemed to stick in reverse. After a circuitous journey of 2.3 billion miles, Galileo arrived at Jupiter, where winds average 250 miles per hour and lightning bolts are 100 to 1000 times more powerful than on Earth. A massive storm on Jupiter has lasted for centuries, manifesting itself as a mysterious Great Red Spot on the planet's surface.

When Galileo was poised at 130,000 miles above the surface of Jupiter, two robot crafts were hurled into the planet's atmosphere. At almost 50 times the velocity of a high-powered rifle bullet, the crafts descended into the planet's ammonia clouds and signalled back to their mother ship, which in turn signalled scientists at NASA's Jet Propulsion Laboratory in Pasadena, California. The return signals were received on 8 December 1995. The smaller crafts were to measure the dust storms that whip around in Jupiter's magnetic field and are produced by its rapid ten-hour rotation. They also sampled the charged sulphur and oxygen atoms trapped in Jupiter's ionosphere. The two probes were expected to burn up after about 3.5 hours, while the mother ship was to orbit the planet 11 times.

Whilst the information sent back to Earth was interesting, it also cost almost $2 billion.[31] With the right combination of community involvement and good government policy, $2 billion could provide good accommodation for 20 million inhabitants of the world's worst slums. Unprecedented cutbacks in welfare and social security were being made in the US and other Western nations at that time.

Astronomers became quite excited when nature joined in their quest for information about the immense energy in Jupiter's atmosphere. The shattered comet Shoemaker-Levy slammed into the planet in July 1994. Shoemaker-Levy was enormous and some of its particles were the size of small mountains. Galileo, then still 150 million miles away, was nevertheless able to send information about the impact back to Earth. At the McDonald Observatory in Fort Davis, Texas, astronomers were 'running around like giddy little kids because [they] can see the structure of the spots. It's so much fun to watch Jupiter change before your eyes.'[32] I find this indicative of our failure to realise that planets are just one part of a very wonderful interactive whole; a collision on one planet can reverberate throughout the whole solar system. As a Native American saying goes, 'all things are connected'.

The fragmented comet impacted in Jupiter's southern magnetic field, causing charged particles to accelerate northward and generating thousands of volts of electricity. Scientists did not expect the effects to be at the pole opposite to the collision area. These 'surprises' only serve to show how little science really understands about atmospheric processes.

Ulysses Mission (1990)

Galileo was the first large spacecraft to be powered by two plutonium RTGs, and it was launched after a decade of citizen protest. Despite this, immediately after its launch NASA announced plans to launch another such mission, Ulysses, within a year and more missions to follow; the launching of plutonium into space was to become a routine activity.

In preparation for Ulysses, NASA prepared a Draft Environmental Impact Statement (DEIS), dubbed by many scientists as by and large a misleading document, making claims to justify a decision already made.[33] NASA announced that it would be using one RTG fuelled with plutonium fuel, and in a notice in the US Federal Register stated that it would not consider any alternative that might cause delay to its programme 'no matter what the public comments might be'.

Although NASA invited public comments on its plan, the key documents on which the DEIS was based were not easily attainable by independent scientists, especially within the time limits specified. The process also highlights the difficulty in rousing citizen participation in decisions involving atmospheric research. In the case of an Earth-based project such as a military testing ground or a depot for weapons storage, there would be public meetings, comments by local scientists and physicians, citizen oversight committees, media investigations, environmental hearings, and other potentially enlightening scrutiny. Space projects, on the other hand, are not recognised as being in anyone's 'back yard' so the dangers seem less immediate. Even when an accidental shower of plutonium occurs, it is invisible and has no taste or sound, so it is difficult to arouse people's concern. This situation is compounded by the military secrecy surrounding its plans for space.

Ulysses is designed to carry 24 pounds of plutonium. In its experiments with the effect of plutonium on beagle hounds, the US government has never found a dose small enough that it does not cause lung cancer. If Ulysses had exploded, the plutonium would have been dispersed over a large geographical area. NASA put the probability of such an accident at 1 in 10 million. As we have seen, such statistics aren't reassuring.

Cassini Mission (1997)

In October 1997, NASA launched a rocket bound for Saturn. This rocket, named Cassini, contained three RTGs powered by 72.3 pounds of plutonium 238, which is about 280 times as deadly as the more common plutonium 239. This rocket had a complicated route: first a fly-by of Venus, followed by a fly-by of Earth, and then on to Saturn. A 'fly-by' means that the rocket enters the atmosphere and is given a push by the planetary gravity and rotation, changing its direction and increasing its speed. Such manoeuvres pose a serious danger of rocket burn-up with dispersal of the plutonium.

NASA began developing Cassini in 1992, and it is the most ambitious project to date, costing an estimated $3.4 billion. In April of 1997, Cassini was transported in secrecy to the Kennedy Space Center in Florida, and in late August 1997 it was moved to a securely guarded launch pad at Cape Canaveral Air Force Station and loaded onto a Titan 4B rocket. The base went on 'Threat con Alpha' alert, military parlance for heightened security conditions to counteract possible terrorism.

While a number of scientists and activists spoke out against the Cassini flight, one of the most impressive speakers was a former NASA employee. Alan Kohn, now retired after 30 years service with NASA, was the emergency preparedness officer for both the Galileo and Ulysses missions and a member of the Radiological Emergency Force Group. He finally spoke out in 1997 in order to warn the public of the dangers of the Cassini launch.[34]

In a speech given outside the gates of the Cape Canaveral Air Force Station, Kohn said that he had been told 'his job was cosmetic, and in case disaster, the unlikely event of disaster, would take place' he could take all protective measures – which did not exist. According to Kohn: 'The only measure I could have taken at that time, of course, would have been to wet my pants!' Kohn went on to say that he was told to 'lay off, keep a low profile, don't let the public know, above all don't let the protest groups know that there is any danger at all'. Kohn said in his speech:

> I disobeyed orders. I provided that all buildings should be turned into fallout shelters, that air conditioning be shut off, that buildings be sealed, the doors be sealed, that people who were going to work outside would be put in bunny suits and given gas masks with HEPA filters [for radiation protection]. I provided washdowns. I told them no visitors [for the launchings]. They brought visitors anyway. And by the way, in the mission control center when I said no visitors, I got an ovation from the people [the NASA employees] ... The people applauded me because they agreed with me. They didn't agree with me publicly but the applause was enough to show me that on the government side of those fences, there are a lot of people who agree with you [citizen protesters] but out of misguided loyalty they don't have freedom

– they think they don't have freedom – to speak out. I disagree. The first loyalty is to the public. The first loyalty is to the taxpayer. The first loyalty is to each other, to our own families.

A second NASA employee, 56-year-old James Ream, was also worried about possible plutonium release, so he joined anti-Cassini demonstrations on 24–31 July. After attending the protests, Ream said that he was questioned by Kennedy Space Center security officials and asked to sign a letter promising that 'he would not aid protesters who might try to infiltrate Cape Canaveral to disrupt the launch'. On 26 August, Ream appeared before the Titusville City Council (the municipality closest to the launch site), and asked members to pass a resolution against the launch. Ream was suspended without pay from 11–12 September, in what NASA claimed was an 'unrelated work infraction'. He had worked for NASA since 1966, had no reprimands in his file and had never been suspended before, according to Ken Aguilar, director of personnel for the Kennedy Space Center.[35]

Launching took place on 6 October 1997 and in August 1999 there was an Earth fly-by. Fortunately, there was no disaster. The total journey to Saturn will take seven years, and if all goes according to plan, the rocket will orbit Saturn for four years.

European researchers, under contract with the European Space Agency, undertook a study to see if the Cassini mission could have carried out its proposed goals using a benign solar energy source. The Europeans have developed high-efficiency silicon solar cells that show promise for use in deep-space missions. The study demonstrated that solar energy could indeed be used for interplanetary travel but the US stated its preference for the RTG. It is thought that this is because RTGs can provide surplus energy that could be used to power weapons in a conflict situation.

The US has sent up 24 rockets with nuclear power capability, and three have exploded. The Russians have sent up 39, six of which have been destroyed by accidents.[36] None of these carried as much plutonium as the Cassini rocket. Scientists have calculated that as many as 20 million people could have developed lung cancer if this had exploded near a populated area.[37]

THE QUESTION OF RESEARCH

The space programme has been costly and it has been dangerous. What is equally disturbing is that all of the atmospheric research described in this chapter has been undertaken by the military and, because of the secrecy shielding military research, it is not always easy for the public to understand the possible consequences. But what of the value of research – hasn't the space programme led to wonderful discoveries about the solar system in which we live?

I would liken society's dependence on the military to a family in which one partner is addicted to something and claims a large proportion of money and resources for feeding the addiction. The rest of the family feels dependent and puts up with a lack of resources rather than disturb the addicted person, who perhaps provides some security in an insecure world. They may also find it hard to imagine that they could survive without this dependency.

I believe we do not need to live with this military addiction in order to progress in the sciences in a positive way. The problem with military research is that it goes in only one direction: the military needs civilian support but civilians have no need for the military. For example, no one would mine uranium if it were only to be used for bombs, and no one would teach astrophysics if it were only to be used for war. Military research is therefore dependent on finding a civilian front industry that will attract civilian, and especially university, cooperation and funding. Such cutting-edge research attracts the brightest young people, offering them exciting challenges and good pay. Researchers are often not even aware of military interest in their research. This 'brain drain' from the civilian economy may be depriving us of those who could resolve the most serious survival problems now facing the biosphere. Although humans have proved themselves capable of serious damage to the Earth, they also hold the key to healing.

As the line between military and civilian research becomes blurred, it can be difficult to identify just when we have strayed from observation and controlled experimentation into manipulation and dangerous practice. At what point is experimentation safe if there are no controls as to what that research will be put to? And do we really learn from past mistakes?

What becomes apparent from the last thirty years is that space experiments are not just about exciting scientific exploration. Space is the next battlefield. According to General Joseph W. Ashy, commander-in-chief of the US Unified Space Command:

We'll expand into these two missions (space control and space force application) because they will become increasingly important. We will engage terrestrial targets someday – ships, airplanes, land targets – from space. We will engage targets in space, from space. And this command will engage quickly; [the missions] are already assigned, and we've written the concepts of operations. We will engage re-entry vehicles in the medium of space with a ballistic missile defense system of North America. It's politically sensitive, but it's going to happen. Some people don't want to hear this, and it sure isn't in vogue... but – absolutely – we're going to fight in space.'[38]

CHAPTER 3
MILITARY PLANS FOR SPACE

Taking war into space is not a new idea. In the 1960s the Soviets had an orbital weapon called a 'killer satellite'. It was supposed to use a radar-guided, two-orbit profile (that is, it needed two orbits around the Earth) to identify and lock on to its target. Since the two-orbit timing was slow, attempts to create a one-orbit 'kill' using infrared guidance were tried but were notably unsuccessful. The Soviets also had an orbital weapon known in the US as a FOBS – fractional orbit bombardment system. The idea was to place a hydrogen bomb in low Earth orbit so that it could be quickly launched to a ground target if needed. The system was secretly tested from 1966 to 1970. The civilian population never knew that thermonuclear bombs, about a thousand times more powerful than the Hiroshima bomb, were orbiting above their heads.[1] FOBS are not thought to be operational now but the Soviet government has revealed that it has 18 FOBS launchers in their inventory at Tyuratam. Following the economic collapse of the country, the entire Soviet space programme, including the FOBS, may end up for sale to the highest bidder.

With the clampdown on nuclear testing of the atmosphere, the US military sought a new umbrella under which to continue their research and experimentation with the ionosphere. They also needed civilian support and funding, especially from university programmes. In the late 1960s the public, disillusioned with nuclear energy and worried about acid rain, was eager for alternative forms of energy. The solar power satellite project emerged as a programme that would simultaneously win public approval and enable the military to explore their vision of war in space.

MISSILE DEFENCE SYSTEMS

SPS: Solar Power Satellite Project (1968)

In 1968 the US government proposed a system of satellites that would capture energy from the sun and beam it to Earth for domestic use. Each satellite was to be the size of Manhattan Island and was to be placed in geostationary Earth orbit (GEO). A satellite in GEO appears to be always at the same point above the Earth's surface because it takes exactly 24 hours to circle the planet, therefore matching the Earth's rotation. GEO is normally about 40,000 kilometres (25,000 miles) above the Earth, in the Van Allen belt region. Around 60 such satellites were needed and they were to be built in orbit over a 30-year period.

It was planned that the satellites would intercept solar radiation using solar cells, and then transmit the energy via a microwave beam to receiving antennas, called rectennas, on Earth.[2] The rectenna sites, each capable of generating five gigawatts of electricity, were expected to take up to 145 square kilometres of land, and would preclude habitation by humans, animals or even vegetation. Plans for controlling wildlife and birds were sketchy, depending, apparently, on letting them die if they wandered too close to a rectenna or flew through a microwave beam. The system also envisaged use of 'modular rectennas', mobile units that could be used instead of oil generators when the military needed ample electricity in remote areas.

The construction phase of the project was expected to require three to five daily launchings and re-entries of large transportation rockets in order to transfer workers and equipment to and from orbit. At the time, the space shuttle was only on the drawing board, but it was recognised that using a rocket once and then discarding it was financially prohibitive.

The US Congress mandated the Department of Energy and NASA to prepare an Environmental Impact Assessment for this project, to be completed by June 1980. The assessment alone cost $25 million. In fact, everything about this project was expensive. In 1968, it was estimated that building the 60 satellites required would cost between $500 and $800 billion. The system would provide around 10 per cent of US energy needs by the year 2025 but at a cost of $3000 per kilowatt. Even nuclear power only cost $1100 per kilowatt at the time.

The need for such a high-tech, high-cost form of energy escaped most ordinary people, who were more used to the idea of harnessing solar energy through ground-based solar panels. At the time, about 70 per cent of heating and air conditioning could have been generated directly by such a method, absorbing warmth from the sun and storing it for use when required. This would dramatically reduce the need for electricity, making the use of benign technologies such as wind energy more feasible. To environmentalists the low-tech solution to the energy crisis made more sense than the plan to put expensive satellites in the Van Allen belts.

Early review of the Solar Power Satellite (SPS) Project began in 1978, and I was on one of the review panels. I knew very little about the military experiments at the time, and almost nothing about the ionosphere and its complex structure. When I went through school, very little of today's astrophysics was known. Certainly, the politicians of the time were in a similar state of ignorance, and I do not think that the other civilian reviewers on the SPS Environmental Impact Assessment panel knew much more. Most of the nuclear events and military experiments were kept very quiet or, if reported, were claimed as military 'successes' in the nuclear or space race. Hence, we approached this task with open and unbiased minds.

Although it was proposed as an energy programme, it was clear that the SPS had significant military applications. One of the most significant, first pointed out by Michael J. Ozeroff, one of the reviewers on my panel, was the possibility of developing a satellite-borne beam weapon for anti-ballistic missile use. It was speculated at that time that a ground-based high-energy laser beam could function as a thermal weapon to disable or destroy incoming enemy missiles. Therefore imaging it operating from a space platform was reasonable. There was some discussion of electron weapon beams, through the use of a laser beam to 'clear' a path for electrons to follow, but this was all highly theoretical. The height of the platforms, in the Van Allen belt, made this scenario credible.

Michael Ozeroff noted other potential military uses for the SPS, including surveillance and early warning of a hostile attack. The satellites were to be in geosynchronous orbits, each providing an excellent vantage point from which an entire hemisphere could be surveyed continuously. As early as 1960, the US Carona programme had placed a spy satellite in space to photograph Soviet missile bases. This was followed by the US Keyhole series of satellites (still classified), which are said to be able to detect an object the size of a car licence plate from an altitude of 160 km (100 miles) or more.[3]

At the time, security restrictions meant that we were allowed to discuss theoretical possibilities but could not connect these with any specific military programmes, either in existence or at the planning stage. Even so, some potential functions were obvious: the SPS platform could be used for communications, jamming, navigation, meteorological-geological-geographic surveying, and extra-low frequency links to submarines.[4] 'The SPS beam could possibly be redirected to jam or destroy nearby space satellite systems; and certainly terrestrial communications, both tactical and strategic, could be affected in a major way.'[5]

The SPS was also described by Ozeroff and other reviewers as a potential psychological and anti-personnel weapon. It could be used to cause panic by knocking out all electronic devices (computers, traffic signals, televisions, radios and so on), thereby paralysing a city; and if two main microwave beams were redirected away from their rectennas to intersect near enemy personnel, the SPS could function like a microwave oven – it could kill people and spare buildings, just as a microwave oven cooks food but leaves a paper plate untouched. It might also be possible to transmit high enough ultraviolet energy to Earth to ignite combustible materials, like a dry forest or an underground oil deposit. Laser beam power relays could be made from the SPS satellite to other satellites or platforms, for example to aircraft, for military purposes. One application might be a laser-powered turbofan engine that would receive the laser beam directly into its combustion chamber, producing the required high-temperature gas for its cruising operation. This would allow the aircraft unlimited on-station cruise time.

President Carter approved the SPS Project and gave it the go-ahead, in spite of the reservations which many reviewers, myself included, expressed. The cost of the SPS scheme was two to three times greater than the budget of the whole US Department of Energy, and the projected cost of the electricity it would produce was well above that of most conventional energy sources. This reality eventually 'killed' the proposal in the US Congress, and funding was denied. At the time, I approached the United Nations Committee on Disarmament about the plan, but was told that as long as it was called solar energy, it could not be considered a weapons project.

Strategic Defense Initiative – Star Wars

The same project resurfaced in the US under President Reagan in March 1983. It was moved to the much larger budget of the Department of Defense and was called the Strategic Defense Initiative (SDI), but is more widely known as 'Star Wars'. It must be remembered that military programmes do not originate in the minds of elected officials, but in the strategic planning departments of the army, navy and air force. Military strategists will continue to push for what they want even if their proposals are rejected the first time round.

Reagan's Star Wars envisioned a layered defence against incoming intercontinental ballistic missiles, counting on the fact they would require about 30 minutes from lift-off in the Soviet Union to touch down in the USA. For the system to work, it would require almost immediate detection of the hostile missile launch, and this would be achieved by using surveillance satellites. The enemy missile would be attacked by either satellite or ground-based missiles, hopefully striking it on lift-off before its warheads could be deployed. Then the surviving missiles, if there were any, would be attacked by X-ray and particle-beam weapons. All of this would be managed by supercomputers whose infinitely complex programs would be written by other supercomputers, with no human intervention allowed or required. The cost of building this defence system was projected to be anywhere between $100 and $1000 billion.

The Star Wars programme was opposed by scientists on both sides of the Atlantic, some because they believed it would never work, some because it was too expensive, and some on the basis that it would violate the Anti-Ballistic Missile Treaty agreed by both the US and the USSR in 1972 – a treaty that is viewed as fundamental to nuclear arms control. The treaty limited both countries to no more than 100 anti-ballistic missile (ABM) weapons apiece, with each side agreeing to deploy its missiles around only one site (Moscow and Washington DC).

ABM systems included nuclear-armed warheads, large radars for identifying incoming missiles, tracking radars, and early-warning networks. At the time, the Soviets had two types of ABM: Gorgon missiles designed to intercept attacking missiles in the Earth's upper atmosphere, and Gazelle missiles for short-range attacks within the Earth's lower atmosphere. The US had the Spartan missile with a reaction time of 30 seconds, which could intercept missiles up to 725 km (450 miles) away and at heights of 565 km (350 miles). If the Spartan failed to destroy the missile, the Sprint short-range missile would be deployed to intercept it as it descended towards its target.

The theory was that a first strike by either nation would be unlikely because neither could defend its whole territory against a retaliatory strike. Put simply, you don't attack someone if you know that you can't defend yourself when they fight back.

Little diplomatic conversation had prepared America's European colleagues for the Star Wars scheme, which was seen as a further escalation of the nuclear arms race and a break from previous nuclear arms strategy. Both NATO allies and the anti-nuclear movements spoke out against it, stressing not only that it was unwise and broke the terms of the 1972 agreement, but also that it was likely to be impossible. Yet, in spite of severe international criticism, the US began work on this programme, spending about $30 billion between 1983 and 1993.

Missile-intercept technology was tested during the Gulf War. As was noted in Chapter 1, the Patriot basically failed this test, although this was not generally admitted until the war was over. One of the Gulf War commanders used the phrase 'maximizing our potential for violence' to describe his goal of improving the accuracy of each missile so that each shot counted. In fact, the accuracy was shockingly poor.[6] At a factory that produced the Patriot missile,

President Bush proudly stated: 'Forty-two Scuds engaged, forty-one intercepted!'[7] By the end of the war the count was actually 85 Iraqi Scud missiles fired – 25 veered off into either the Israeli or Saudi desert or landed in water, 55 hit civilian targets in Israel, and only 5 were intercepted by the Patriot missile.[8]

It has also been estimated that in the Gulf War 77 per cent of all the damage to American combat vehicles was due to 'friendly fire'. Also on 27 February 1991, an American A-10 jet mistakenly shot at two British Warrior infantry vehicles, killing nine people and injuring eleven. This was called a failure in 'intelligence', and there were strong demands by all sides for precise information, delivered immediately to all parties involved. It is obvious that this high-tech electronic warfare made all ground personnel vulnerable, and it does not require much imagination to extrapolate the 'mistakes' of manned electronic warfare to the unmanned version envisaged in the Star Wars programme.

Anonymous disillusioned researchers in the US alleged that the US Department of Defense had falsified data on missile interception tests and even rigged some to fail in order to gain Congressional support and money for further research and development.[9] Casper Weinberg, who was Secretary of Defense under President Reagan, denied these claims. He subsequently stated that Star Wars was designed only to deceive the USSR so that they would spend their scarce currency on space-based defences. Whatever the reality behind these claims may be, it is important for the reader to understand that whether these projects were proposed as solar energy or missile shields, they did not stop once the Cold War was over.

Ballistic Missile Defense Organization

In May 1993 the Star Wars project was officially ended and the funds still in the budget were transferred to the Ballistic Missile Defense Organization, which appears to be the third official name for the same project. This new agency was to concentrate on theatre or tactical missile defence (TMD) – countering low-flying short-range missiles rather than intercontinental ballistic missiles. In July 1993, President Clinton denounced Star Wars as a violation of the

1972 ABM Treaty, yet in August 1995 the US Congress funded a programme to establish a network of ground-based missiles that would be guided by space-based sensors designed to locate incoming long-range missiles.

It is difficult for the outsider to understand how these two programmes really differ from the Star Wars scenario, but it seems that they too may violate the 1972 ABM Treaty, which limited not only the number of ABMs but also the number of sites. The treaty is constantly being reinterpreted by military planners according to the defence needs of the time and, as we shall see, the definitions contained within it are presently being stretched beyond all recognition.

Laser defence

Whatever their name, the missile programmes described above can be divided into the same basic components: surveillance, tracking, and missile interception from space, land, sea and air. Each military branch puts forward its own requirements and competes for funding and new weapons.

Increasingly, modern weapons technology relies on the use of lasers for both guidance systems and for attack. LASER stands for light amplification by stimulated emission of radiation. An atom usually exists in what is called its ground state, which is the state in which it has the least energy. By introducing more energy, the atom becomes 'excited', meaning that the outer electrons begin to move in a higher orbit than usual around the nucleus of the atom. If you take a cell containing a large number of the same atoms all in their ground state and energise the whole cell, it is called a population inversion. Normally a population inversion cannot be maintained, because the electrons soon lose energy and fall back into their natural ground state. However, if you can excite the electrons into a still higher state, they will fall back to the first 'step up' and stay there for a short time. When one cell then drops back to ground state, it will trigger the whole population to fall back simultaneously. The excess energy is released in the form of photons (light), packets of electromagnetic radiation. In a laser, these photons all have the same energy, so the light emitted is monochromatic (all ultraviolet or all X-rays, for example). The waves emitted are all in phase (in synchronous movement), forming a single wave. This enables precise focusing of the beam, which can be made very powerful, travelling from Earth to the moon or cutting through metal. The process of exciting the cells is called pumping and the source of energy used is called the pump. Some pumps are: camera flash, sunlight, other lasers, electrical discharges, chemical reactions and, for the X-ray laser, nuclear explosions. If the pump is applied in a rhythmic way, you can produce a pulsed laser. The medium or cell population used for producing a laser can be almost anything: a gas, a solid, liquids, and ionised plasma.[10]

US funding for Brilliant Eyes, the military jargon for the navy's strategic missile and tracking systems, the anti-satellite technologies programme, and the nautical high energy laser, was authorised as an 'add-on' to the National Missile Defense request of the Clinton administration in 1996, purportedly 'to deal with terrorist rockets'. The total package went for close to $1 billion.[11]

The US Senate also added on $70 million for the air force's space-based laser missile, even though this was opposed by the Office of Budget Management. Lasers were first used by aircraft to guide bombs to their target – sensors in the bomb's nose locked onto the reflections of the laser beam. Today's laser-guided bombs (LGBs or 'smart bombs') are marketed by Texas Instruments in the US and Tricon in the UK, and they are called Paveway III. The US and UK both implemented Paveway II in their aircraft in the 1970s.[12] As laser technology becomes more advanced lasers are also being installed in aircraft as beam weapons. An article in *Air Force News* describes how a Boeing 747-400 aircraft is being adapted so that it has a turret 'from which a beam of laser light will emanate to destroy Scud-like missiles hundreds of miles away'.[13] This modification is part of a research project that aims to establish an airborne laser fleet which could participate in battle:

Two Attack Lasers would be flying around the clock, orbiting at about 40,000 feet, providing defense against attacking missiles. If the enemy were to launch a theater missile, Attack Laser would detect the booster while it is still powered as it emerges through the clouds. The Attack laser would then destroy the missile, with resulting debris falling back on enemy territory.[14]

The army has also been developing its laser capabilities. For some time a ground-based laser has been operating at the White Sands military base, New Mexico, and it has been tested on 'a variety of things' according to Kenneth Bacon, Defense Spokesperson for the Pentagon.[15] The White Sands laser, called MIRACL or Mid Infrared Advanced Chemical Laser, uses deuterium fluoride and helium and is considered to be the most powerful laser in the US, capable of destroying satellites and disrupting 'enemy' communication and surveillance. Bacon confirmed that the US military carried out a test on a US air force satellite that had reached the end of its useful life in September 1997. The stated purpose of this laser is 'to protect assets in space and to control space to the extent necessary for our national security interests'. Tests on portable Earth-based laser weapons also began in 1998.

Radar systems

RADAR stands for radio detection and ranging. Radar uses a transmitter that generates radio waves, and when an aeroplane or missile enters these waves it scatters a small amount of the energy back to the receiving antenna. This weak signal can be amplified and displayed on a screen. Because radio waves travel at a known speed of 300,000 km per second (or 186,000 miles per second), the distance to the object can be calculated by measuring the time between sending out the radio wave and its return.

A geographically widespread chain of very large radars was built across northern Alaska and Canada in 1962 to detect any incoming missiles from the USSR. In 1985 all of the Distant Early Warning (DEW line) radars were replaced with large phased-array radar systems using 100 solid state modules (transmitters) which are capable of picking up signals and tracking a target at a distance of 5560 km (3455 miles). This was supplemented by an over-the-horizon radar in a system called NADGE – NATO Air Defense Ground Environment. Over the horizon radars tend to be used to detect incoming enemy bombers, whilst satellites are normally used for spotting missiles.

A new Ballistic Missile Early Warning System (BMEWS) is now operative and widely dispersed geographically. This system is capable of detecting the position, velocity, launch site, trajectory, impact point and impact time of an incoming missile.

Stretching the Boundaries

When the Ballistic Missile Defense system is completed, it is expected to have two major components: a National Missile Defense, to protect North America, and a Theater Missile Defense, intended to protect American troops from missile attacks wherever they are stationed. This programme continues to experience repeated and serious hardware failures.[16]

In 1993, the United States initiated talks with Russia in order to clarify the demarcation between defences permitted and those prohibited by the 1972 ABM Treaty so that it could legally proceed with its research. The new Theater High Altitude Area Defense (THAAD) and Navy Theater Wide systems blurred the definition of 'theatre defence' because both were designed to be mobile and could intercept much longer-range missiles than previous systems. They could, in theory, be deployed to defend the United States from strategic missiles. After four years of often deadlocked negotiations, Russia agreed that testing of THAAD could proceed, but no real agreement on amending the treaty's terms was reached.

Again, in the summer of 2000, President Clinton attempted to renegotiate the treaty with Russian President Vladimir Putin. According to Reuters on 2 June 2000, Washington wanted an integrated defence system that could be deployed by 2005, to defend all of the 50 states against incoming warheads from 'rogue nations' such as North Korea and Iran. It was estimated that North Korea might pose a nuclear threat by 2005. The proposed system seemed to involve a 'complex system of targeting radars, interceptor missiles and high-speed computers' – a description strikingly similar to the Star Wars concept. The funding for this programme must be enormous, as each month of delay is estimated to cost $124 million.

President Clinton was under pressure to come up with a result because he had to decide whether to authorise the first stages of a radar-tracking system on Shemya Island at the western end of Alaska's Aleutian Island chain. The Pentagon warned that any delay in the construction schedule would jeopardise its ability to meet the

2005 deadline. Luckily for President Clinton, on 14 June 2000, lawyers said that the clearing of land and laying of foundations would not violate the 1972 treaty, thereby leaving the knotty question of whether to withdraw from the treaty altogether for the next administration (either side can withdraw if they give six months' notice). 'Asked today how the threshold for violating the treaty for most of Mr Clinton's tenure – pouring concrete – could suddenly become the basis for a broader interpretation permitting construction, a senior Pentagon official said "Better lawyers".' This seems a remarkably blasé attitude towards a treaty which has been the foundation of nuclear arms control.[17]

The 2005 scheme is extremely controversial. A variety of prominent scientists and former Clinton administration officials have urged him to defer his decision. A classified report by a Pentagon-appointed panel of experts, headed by Larry Welch, a retired four-star general and former air force chief of staff, raised numerous issues – problems with booster rockets, concern that the timetable is unrealistic and does not allow for adequate testing, and doubts about whether the interceptor missiles can distinguish enemy missiles from decoys.[18] A report from the General Accounting Office warned that the plans were based on 'uncertain assessments of the potential threats' and concluded that 'it will be difficult to know whether the missile shield will function properly during an attack'.[19] Bearing in mind the abject failure of the Patriot missile during the Gulf War and the deaths caused by friendly fire, this is an extremely worrying consideration. Even more troubling is the fact that a US shield could spark off an arms race with Asia.

The components of the original SPS proposal have now become so fragmented that what was once a comprehensive system is now being judged and funded on a piece-by-piece basis. Each piece considered by itself seems not to be threatening. For example, one of the first components to be funded and completed in the early 1980s was the Space Shuttle, a retrievable rocket which could transport workers and materials to and from space. From the vantage point of someone who reviewed the SPS project in 1978–80, it appears to be a jigsaw puzzle slowly beginning to take shape. I do not think this view qualifies as a 'conspiracy theory', it is rather a reflection of the persistence of research ideas over large time frames. As science makes new discoveries, so we will see new weapons emerging and being incorporated into this race to dominate space.

DESIGNING WEAPONS AND COMMUNICATION SYSTEMS FOR THE NEXT WAR

Space Shields

The Star Wars project and the Ballistic Missile Defense programme both envisaged putting up 'shields' of protection over regions of the planet through the use of ABM technology. One way we might see this concept evolving in the future is thought to be through the use of plasma shields.

As discussed in Chapter 2, plasmas are a superheated gas and they occur naturally in the ionosphere. We can see the effect of plasma on an object when we consider how the ionised layer of the Earth's atmosphere burns up the space debris and meteorites that enter it. It is not really friction with the more dense atmosphere which causes the extreme temperature, but the impact of the spacecraft itself, which compresses the highly active plasma, causing its temperature to increase dramatically – it can even temporarily reach the temperature of the surface of the sun. The Space Shuttle has insulation tiles on its surface to protect it from this heat.

Plasma exists briefly in the lower atmosphere, the troposphere, when lightning strikes. One lightning strike from a cloud to Earth consists on average of four strokes in rapid succession. At all times, lightning is hitting the Earth somewhere, charging it negatively with respect to the ionosphere by roughly 200,000 volts. If lightning ceased even for an hour, Earth would discharge its stored electricity, causing untold damage.

A phenomenon associated with lightning is ball or globe lightning. This is a glowing, floating, stable ball of light, occurring at times of intense electrical activity in the atmosphere. On contact, these balls release large amounts of energy. Ball lightning occurs near to the ground during thunderstorms and may be red, orange or yellow. It is accompanied by a hissing sound and has a distinct odour. The causes of ball lightning are unknown but speculated causes include air or gas behaving abnormally; high-density plasma; an air vortex containing luminous gases (a vortex is a whirling phenomenon like a miniature whirlwind); and microwave radiation within a plasma shell.

Scientists have theorised that a microwave generator could be used to fire a plasmoid, a blob of plasma not unlike ball lightning, into the path of an incoming missile, its warhead, or an aircraft. The theory is that as a missile passes through the fireball, its electronics and navigational systems are disabled. Electromagnetic energy also interferes with the isotopes of a nuclear warhead, effectively disarming the weapon.

The Banjawarn Event, 28 May 1993

On 1 June 1993, the *Kalgoorlie Miner*, a newspaper in Western Australia, reported that a meteor fireball flying from south to north between Leonora and Laverton had been seen by several observers on 28 May. This sighting was followed by an earthquake, measuring about 4 on the Richter scale at 23 different seismic receivers around Western Australia. Ed Paul, a geophysicist who recorded the event, also noted that the quake sheared three-inch steel pipes underground at the Alycia Gold Mine and collapsed underground drives and shafts. This is a significant finding, since in quakes induced by seismic ground waves damage is usually limited to surface building collapse. Ed Paul thought that a nuclear explosion had occurred.

Many observers reported that the fireball passed overhead, making a pulsed roaring noise, similar to a very loud diesel train. After the seismic wave hit, they reported hearing a huge, long, drawn-out explosion similar but not quite the same as a mine blast. Although seismographs have been in place since 1900, there is no record of any previous earthquake in this area, nor do aboriginal people have any memory of such an event. Everyone assumed that a meteor fireball, a bolide, had struck the Australian outback.

The point of probable impact was in the Eastern Goldfields region of Western Australia, a very isolated and scarcely populated semi-desert region. A geologist, Harry Mason, visited the site out of curiosity in May and June 1995 and was surprised to find no sign of an impact crater or ground anomaly. He interviewed as many witnesses as he could, and discovered several facts:

> people heard the fireball well before they saw it; it was a large orange-red spherical fireball with a very small bluish conical tail; the speed was like that of a 747 jet liner; the fireball flew apparently parallel to Earth's curvature in a long 'nap of Earth' arcing trajectory at low altitude (about 2000 metres) over a distance of at least 250 kilometres; the fireball arced down toward the ground and disappeared behind trees or low hills; then a blinding massive high energy burst of blue-white light lit up the night as if it was daylight. Observers could see more than 100 kilometres in every direction at ground level; a red colour flare then shot vertically skywards and a massive seismic ground wave hit the observers; a very loud major explosive blast followed, that was heard over a 250 km by 150 km corridor; minor quake damage was reported as far as 150 km southeast.[20]

There seemed to be ultrasonic or electromagnetic waves which dogs were sensitive to since they went totally berserk, whining and howling while the sky was lit up. Exactly one hour after the first fireball event, there was a second smaller one. Later (the exact time is not known) there was a third fireball that passed over Banjawarn and was reported by truck drivers.

The area in which this event occurred had just been purchased by the Japanese Aum Sinri Kyo 'Supreme Truth' Sect, accused of instigating the 1995 Tokyo subway gas attack. The deal was closed on 23 April 1993, only 35 days before the first fireballs were seen. The deputy leader of AUM initiated the purchase to 'conduct experiments there for the benefit of mankind'. It cannot be shown whether or not these experiments had any connection with the fireballs. The chief lawyer of the US Senate Inquiry into the AUM sect informed Harry Mason of the AUM's great interest in electromagnetic weapons and their power to induce earthquakes. AUM sect personnel were at Banjawarn the night of the strange event.[21]

Electromagnetic weapons are believed to be variants on designs proposed by Nikola Tesla in 1908 and have the ability to 'transmit explosive, and other effects such as earthquake induction, across intercontinental distances to any selected target site on the globe, with force levels equivalent to major nuclear explosions.'[22]

Nikola Tesla was a Serbian-American, born in Croatia in 1856. He emigrated to the US in 1884 and worked under Thomas Edison. He and Edison disagreed on whether commercial exploitation of the newly discovered electricity should be based on direct current or alternating current systems. In 1888, Tesla demonstrated that if two wire coils were placed at right angles to each other, and they were both supplied with alternating electrical currents out of phase with one another, they could be used to make magnetic fields rotate. This basic electric motor design was purchased and promoted by George Westinghouse, who recognised its value for household appliances. Tesla influenced the decision by American power plants to opt for alternating current rather than direct current for electrical transmission and domestic use. He did brilliant work in high-voltage electricity and wireless communication, and he had ambitious plans for using electromagnetic power to create weapons. He died in 1943 before he could design and build the many devices he had conceived.[23]

Since May 1993 there have been thousands of sightings of aerial fireballs and associated light-energy emissions in Australia. One event was observed by about 500,000 people in Perth, who were awakened by the violence of the explosive seismic wave. These events have not been widely covered by the international media and locally people have been told that they are all caused by meteors. However, a meteor does not have the comparatively slow speed observed in these sightings, nor does it follow an arcing trajectory. Moreover, after a meteor hits there is a crater and fragments can be recovered. There are no craters and no fragments have been found. The projected global trajectories of the fireballs pass near four military complexes: Showa and Mizuho in Japan, and Molodezhnaya and Novolazarevskaya in Russia. It is suspected that the Kamchatka peninsula, in Siberia, may be one of a worldwide series of former Soviet electromagnetic weapons transmitter complexes. It was over the Kamchatka peninsula that a Pan Am commercial plane, thought to be spying, was shot down by the Russians on 31 August 1983.

A United States Senate inquiry into the 28 May fireball over Western Australia took seriously the theory that Russia, known to have investigated Tesla physics, was testing a new superweapon capable of inducing earthquakes a hemisphere away. Some Japanese investigative reporters and Australian and American researchers also believe that the Russians have had Tesla-type weapons since 1963.[24] The Senate hearing consulted the US Incorporated Research Institutions for Seismology (IRIS), and while the latter accepted the possibility that the fireball was a clandestine event, they decided that it was more likely to be a meteor strike. IRIS did, however, want to make further inquiries. Other American scientists believe that the fireball was a 'Tesla shield', part of an anti-ballistic missile defence system.

Other Electromagnetic Weapons

Military tank designers have admitted that current tank technology has reached its limits. If tanks become larger and heavier they also become more vulnerable – an easy target for today's precision weapons. Britain's Defence Research Agency (DRA) at Fort Halstead in Kent has been carrying out weapons research since World War II. This installation is so secret that there is no sign for it on the main road and it does not appear on a commercial atlas. It is also surrounded by a high fence and surveillance cameras. In 1982 the British army commissioned Fort Halstead to undertake 'a short sharp look at the possibility of developing a gun powered by electromagnetic force', and more than £10 million over three years was allotted to the programme.

After some disastrous experiments in uncharted territory, with a projectile coming out of a gun sideways, another weapon catching fire, and what the scientists describe as a 'lot of sparks', they finally built an Electromagnetic Gun Laboratory at Kirkcudbright, Scotland, which opened in 1993. 'According to David Hull of the Novel Weapons Division, which has overall responsibility for the gun, it is conceivable that the weapon will be so powerful and so precise that there will be no defence against it. You might need a few of them; it could end tank warfare.'[25] The race to produce the first 'successful' electromagnetic gun is on, but no one is sure what will happen to a projectile fired through the densest part of the Earth's atmosphere at the high speeds contemplated, nor how many times a barrel can endure the pressure.

According to *Defence News*, 13–19 April 1992, the US deployed an electromagnetic pulse weapon (EMP) in Desert Storm, designed to mimic the flash of electricity from a nuclear bomb. A stream of electrons hitting a metal plate can produce a pulsed X-ray or gamma ray and this is capable of knocking out communications over a wide area. The Hermes III electron beam generator is capable of producing 20 trillion watt pulses lasting 20 billionths to 25 billionths of a second and is housed at the Sandia National Laboratory on the Kirkland Air Force Base in the US. Hermes II had been producing electron beams since 1974. These devices were tested during the Gulf War.[26]

HERO

Electronic devices can interact with each other in unexpected ways. The problem is so widespread that passengers in commercial planes are warned to turn off items such as mobile phones at the beginning of each flight. When we consider the proliferation of electromagnetic and electronically controlled weapons, it is clear that this interaction could be lethal.

The military uses electro-explosive devices (EEDs) to activate the fuses on weapons. They rely on an EED to ignite a missile rocket or aircraft motor, to eject a fuel tank, a crew seat or canopy from an aircraft, to drop a bomb, or detonate a warhead. The problem with the EED is that it cannot discern between intentional and unintentional radio signals. Added to this is the fact that electronic components, chemicals and fuel can be ignited by lightning or static electricity. Lightning was the likely cause of the explosion of the Atlas-Centaur rocket on 26 March 1987. The US air force has documented 773 lightning strikes on US air force aircraft between 1969 and 1979, causing seven aircraft losses and two other probable loses. The air force also cited 150 other 'mishaps' including instrument and flight control failure and fuel tank explosion.[27]

Pentagon military analysts have given the name HERO, meaning Hazard of Electromagnetic Radiation to Ordnance, to radio waves, radar, microwaves, electric emission from power plants, power lines, lightning and static electricity. These are all capable of causing conventional, nuclear or chemical weapons accidents, either by affecting the 'explosive payload' (the warhead), the 'weapons platform' (silo, ground launcher, aircraft or ship) or the 'delivery system' (for example the missile or rocket). HERO can accidentally launch missiles, detonate volatile chemicals, and crash aircraft. The US navy HERO officials claim that 25 accidents are HERO proven or suspected, including the explosion of the nuclear Pershing II in 1985 and the gunpowder explosion aboard the *USS Iowa* in April 1989.[28]

Although silk and polyurethane are known to hold in an electrical charge, some old ordnance, packed in these substances before 1945, is blamed for the explosion aboard the *Iowa* in which 47 sailors died. Just before communication with the men loading the gun turret was cut off, one of them yelled: 'friction...static...oh my god.' After the disaster, investigators found an ungrounded wire, and things like bracelets, watches, rings and necklaces on the dead men. These were forbidden in HERO areas, but often young recruits take such regulations lightly. The technical investigation later found that on the day of the accident, an antenna for the WSC-3 satellite communication system was operating approximately 100 feet from the turret that exploded.[29] This antenna creates a powerful field of electromagnetic radiation, and navy regulations call for a distance of at least 213 feet from any explosive material. The *USS Iowa* is a veritable floating arsenal, generally carrying combined firepower of over three million tons of TNT. It was fortunate that the whole vessel did not explode.

HERO is thought to have caused the crash of an F1-11 jet during the US air strike on Libya in 1986. US Air Force Colonel Charles Quisenberry, who was in charge of the operation in which the US not only lost its plane but also bombed friendly embassies and residences, stated quite simply that the accidents were due to weapons 'interfering with each other'.[30] The US over-the-horizon radar, PAVE PAWS (Precision Acquisition Vehicle re-Entry Phased Array Warning System), at Robins Air Force Base near Macon,

Georgia, was found capable of crashing an incoming plane or even launching or detonating the plane's missiles, yet the military failed to shut down or move the radar to a location more remote from the landing strip. Instead they issued 'a finding of no significant impact' in October 1990 and turned the radar off when they knew of a plane landing.

The US Department of Defense reported 20,269 accidental deaths among soldiers and sailors between 1 October 1979 and 20 September 1988.[31] Since investigative procedures commonly omit HERO, it is impossible to calculate how many of these fatalities are due to it. However, in 1989 Senator Sam Nunn in the US ordered an investigation of HERO dangers. John McFall, member of parliament, launched a similar inquiry in Great Britain. In Germany, both the Green Party and the German military have initiated investigations. The Germans became involved when five Blackhawk Army helicopters crashed in Germany, killing all aboard, after they flew near radio transmitters. A German Tornado Fighter bomber also crashed near Munich after flying too close to the Voice of America radio transmitter.

As microelectronics invade every segment of the weapons industry, it is clear that the new high-tech weapons could be vulnerable to the same sort of problems. Ballistic missile defence systems, for example, are to be operated by 'supercomputers' and will launch automatically 'on warning'. This leaves little margin for error. On 25 January 1995, nuclear war was narrowly averted when the Russians mistook a Norwegian rocket launched from the Artic island of Andoya as a US attack on Moscow. Although President Boris Yeltsin issued an attack alert, the order was rescinded twelve minutes into the launch when Russian military advisors noted that the missile was not headed towards Russian territory.[32] In a defence system fully controlled by computers, Hero-type interference could be disastrous.

CYBERWARFARE

On 30 May 2000, General Henry H. Shelton, chairman of the US Joint Chiefs of Staff, released the military's vision for the year 2020. The press release stated:

The overarching focus of JV2020 remains a joint force capable of full spectrum dominance ... pre-eminent in any form of conflict. Four operational concepts – dominant maneuver, precision engagement, focused logistics, and full dimensional protection – continue as the foundation of JV2020.

What emerges from the new military planning is that information technology will be key in the development of future warfare. The US will be able to conduct attacks on foreign computer networks while defending its own system against attack. Strikes will include using deception to 'defend decision-making processes by neutralizing an adversary's perception management and intelligence collection efforts'. Faced with such obscure language, the lay reader can only speculate as to what this might entail. The press release concluded that information warfare will become as important as that 'conducted in the domains of land, air and space'. China's military also has stated that it intends to make information warfare a capability equal in stature to its army, navy and air forces.[33]

The need for sophisticated early-warning systems and intelligence sources is part of this overall plan for 'dominance' and an integrated system of defence.

The Eros Data Center

In the early 1970s, the US Department of the Interior US Geological Survey built the Earth Resources Observation Systems (EROS) Data Center in Sioux Falls, South Dakota. Space Shuttle astronauts took thousands of photographs of the Earth's surface, both in natural colour and using polarising filters, and of objects in space including satellite deployment. According to a NASA brochure:

Critical environmental monitoring sites are photographed repeatedly over time; some have photographic records dating back to the Gemini and Skylab missions (1965–75). Earth-limb pictures taken at sunrise and sunset document changes in the Earth's atmospheric layering. Volcanic activity is monitored in cooperation with the Scientific Event Alert Network (SEAN) of the Smithsonian Institute. Meteorological phenomenon are monitored during Space Shuttle missions. Documentation of hurricanes, thunder storms, squall lines, island cloud wakes, and the jet stream complements meteorological satellite data by offering better resolution and stereoscopic coverage of such phenomena.[34]

This centre has been commercialised since the Land Remote Sensing Commercialization Act of 1984, and is now managed by the Earth Observation Satellite Company (EOSAT) under contract to the US government. The centre houses one of the Department of the Interior's largest computer systems with connections to over 50 terminals in the US, on the federal and state levels, as well as multinational corporations. The Data Center is also linked to the US Geological Survey's National Cartographic Information Center Network. As of 1987, the EROS Data Center held six million aerial photographs of US sites and two million aerial photographs of sites outside the US. This is an invaluable database for civilian, and also military, purposes.

The military has been developing supercomputers to handle its massive data collections for years and a breakthrough in technology will allow the networking of centres like EROS with other military computers. Professor Robert Birge, of Syracuse University in New York State, has invented a data cube, slightly larger than a sugar cube, which can hold twenty gigabytes of information (roughly equivalent in size to 4000 Bibles). Birge expects to expand this to 512 gigabytes of information in a device 1.6 cm by 1.6 cm by 2.7 cm. The key ingredient in this super chip is a form of algae called pond scum, found in San Francisco Bay.

Menwith Hill

Integrated intelligence systems need to 'hear' as well as 'see' and in a futuristic war scenario, it would seems likely that sites such as EROS could be linked to a global network of secret listening posts. One such post is Menwith Hill in the North Yorkshire moors of England. Menwith Hill was set up in 1952 by US presidential decree, with no US Congressional debate, and was first made the subject of public scrutiny by British researchers in the 1970s. Although the researchers used open sources of information, they were arrested under Britain's official secrets legislation and prosecuted in the so called 'ABC' trial.[35]

The Menwith Hill spy base was previously known as the 13th USASA Field Station and is run by a staff of over 1800, mostly Americans, who socialise only infrequently with their British neighbours.[36] The site is managed by the US National Security Agency, and while it originally spied on only Europe, it now covers Europe, North Africa and western Asia. More recently the network expanded globally, intercepting Intelsat satellites which handle most of the world's telephone calls, internet exchanges, e-mail, faxes and telexes. This listening post is linked to a spy network that includes Sugar Grove and Yakima in the US, Waihopai in New Zealand, Geraldton in Australia, and Morwenstow in the UK. Menwith Hill played a crucial part in both the Gulf War and the Kosovo crisis and received the 'Director's Unit Award' for the part it played in Desert Storm.

The existence of the Echelon system, a dictionary sorting method used by Menwith Hill, has been officially confirmed by the Civil Liberties Committee of the European Parliament in a report entitled 'Assessing the Technologies of Political Control'. On 27 March 2000 the European Union took the unusual step of establishing a 'committee of inquiry' to probe this system because it was alleged that the US used intelligence gathered via Echelon to obtain an aircraft deal with Saudi Arabia, beating out its competitor, the European Consortium Airbus.[37]

On 5 July 2000, the *Guardian* newspaper announced that the system was again under fire for alleged industrial espionage. A French public prosecutor, Jean-Pierre Dintilhac, 'ordered the French DST counter-espionage service to build up evidence accusing Washington and London of an attack on the fundamental interests of the nation' after a complaint by Thierry Jean-Pierre, a member of the European Parliament and former judge.

Echelon is designed primarily for non-military targets. The system works by indiscriminately intercepting very large quantities of information and siphoning out what is 'valuable' by finding key words. Five nations share the results – US, UK, Canada, New

Zealand and Australia – with each country supplying 'dictionaries' to the others of keywords, phrases, people and places to 'tag'. Tagged intercepts are then forwarded straight to the requesting country. Whilst there is much valuable information gathered about potential terrorists and criminal activity, there seems to be little accountability as to who is regarded as a legitimate target and how the intelligence can be used.[38]

Control of Cyberspace

While JV2020 (the US's military vision for the year 2000) focuses on information technology as a weapon of war, it is also a powerful tool for those who work for peace and who wish to do away with the secrecy that surrounds military operations. Although NATO determined the military course of events during the Balkan war, it was less successful in controlling cyberspace. Internet communication between Europe, the war zone and North America provided the public with many tales that contradicted the official news reports.

To counter this, President Clinton issued Presidential Decision Directive No. 68 on 30 April 1999, forming a new International Public Information (IPI) group 'designed to influence foreign audiences in support of US foreign policy and to counteract propaganda by enemies of the United States'.[39] The US Information Service, precursor of the IPI, consisting of the Airmobile Fourth Psychological Operations Group (popularly called 'psyops'), was also stationed at Fort Bragg, North Carolina. About 1200 soldiers and officers are commissioned to spread 'selected information' to the media. Major Thomas Collins of the US Information Service has confirmed that psyops personnel have worked in the Cable Network News (CNN) centre in Atlanta, Georgia, as regular employees of CNN. Conceivably, they would have worked on stories during the Kosovo war.[40]

The new buzz words are that information should be 'deconflicted' and 'synchronised', which might be an honest attempt at distributing truth. However, it can also mask an elaborate propaganda operation.

TAKING STOCK AT THE END OF THE 20TH CENTURY

As technology becomes more advanced, the margin for error becomes slimmer. We have created hugely complex systems that are vulnerable to electronic failure, computer malfunction, enemy 'hacking' and electronic interference. Modern weapons are so powerful that any accident could result in huge loss of life and large-scale destruction of the environment. The race to produce bigger, better arms seems to prioritise speed over safety, and again we find ourselves in a situation where weapons are being tested without an accurate view of what they might be doing to us and to the natural world.

Civilian investigation of disasters is severely hampered by ignorance of military technology and secret military experiments. One of the best-known examples of unsolved civilian plane disasters is the downing of TWA flight 800 off Long Island on 17 July 1996. The civilian investigation of this crash identified three possible causes: a missile, a bomb or a mechanical error. After a year, the civilian investigation concluded that the centre gas tank on the jet plane had exploded. It contained gas fumes and was partly empty. The cause of the explosion has not been determined, but is usually said to have been a 'spark'. However, no source of a 'spark' has ever been identified.[41]

TWA flight 800 was a Boeing 747-100, a plane that had accumulated 93,303 flight hours and made 16,869 round-trip flights. The Boeing 747 has the longest flying record of any commercial plane and is considered to be very safe. Flying with partially filled gas tanks is not unusual and the electrical wiring in the Boeing 747 is not near to the centre gas tank. Investigators could not determine what sparked off the explosion, but they ruled out a bomb or missile. There were no bomb fragments found, and the known military vessels near the disaster were thought to be too far away to have reached it with a missile. Therefore this crash is still officially called an unsolved mystery.

There is a very credible account of the downing of TWA flight 800 in a book by James Sanders.[42] According to this account, the plane was brought down by a navy missile experiment that went very wrong.

Secretary of the navy, John H. Dalton, had testified before the US

Senate Armed Services Committee that the navy Cooperative Engagement Capability (CEC) would be undergoing live firing tests beginning in the summer of 1994 and continuing through 1995 and 1996. CEC was the key element in the Advanced Concept Technology Demonstration (sometimes known as Mountain Top) that was tested in Hawaii in February 1996. The system was supposed to be able to distinguish an 'enemy' missile or aircraft among the clutter of civilian aircraft and other decoys. In the tests, a navy missile would intercept and 'kill' a navy drone, or dummy, missile. The system was moving towards combat certification, and therefore funding and production permissions, with increasing levels of realistic testing.

According to Sanders, the final test of this system was being conducted off Long Island on 17 July 1996, the night of the disaster. The drone had been fired, and the navy's radar and target management system, AEGIS[43] computers, were supposed to direct an unarmed missile towards it. The area was chosen because of its complexity, constant flow of commercial traffic and dense ground clutter. The problems may have been compounded by the late take-off of TWA 800, about an hour later than scheduled. The navy test went very wrong. There was heavy electronic jamming in the crowded sky, and the AEGIS-CEC radar-tracking device locked on to the TWA plane rather than the drone, sending the attack missile ripping through the fuselage, several feet below the passenger cabin.

> There was no instant explosion, as the dummy warhead missile sliced through the huge plane as a sheet of paper, depositing a trail of reddish-orange residue in its wake. It roared through the fuselage and exited through the left side of the plane, just forward of the left wing, where it left a hole large enough to walk through . . . [44]

The evidence to support this theory includes a clean entry and exit hole in the forward cabin, 34 certified eyewitness accounts and Federal Aviation Administration radar tapes of a projectile heading towards TWA flight 800, and US government documents confirming naval testing in the area that night. A picture of seats recovered from the plane, with the telltale red-orange stains from missile fuel, can be seen in Sanders' book. There were 212 passengers and 18 crew members aboard the doomed plane. There were no survivors.

This story gives us a frightening view of what could happen in futuristic wars managed by super computers. Yet even if war never occurs, the actual process of weapons development can have detrimental effects on humans and their environment. In the next chapter we will look at how global climate change and freak weather could be symptoms of this military excess.

CHAPTER 4
DOWN-TO-EARTH PROBLEMS WITH STAR WARS

Scientists have always been interested in space and the interdependence between natural Earth systems and the outer layers of the Earth's atmosphere. In particular they are fascinated by the processes that control our weather and climate. In fact, the history of interest in these natural processes dates right back to earliest times because the weather has such a fundamental effect on our ability to survive. Being able to predict weather patterns means we can decide on which the kinds of crops are appropriate, the best times to plant these crops and when to harvest them. It tells us when we might be in danger from extreme weather conditions such as hurricanes, floods and droughts and enables us to take precautionary measures. Understanding of climate influences such things as building codes and urban planning. Understanding atmospheric processes has also been invaluable for shipping and aviation.

Systematic weather records were kept after instruments for measuring atmospheric conditions became available in the seventeenth century. In the US national weather services were first provided by the Army Signal Corps in 1870, again showing the link between the military and early exploration of the atmosphere. In the twentieth century, weather forecasting became a tool in the global marketplace as traders and investors assessed areas of risk and potential growth. In 1977, for example, the price of frozen concentrated orange juice doubled within a matter of days after a huge freeze in Florida.

All of the space activity considered so far has required initial research into the composition of the Earth's atmospheric layers and how they function. For example, to ensure the accuracy of the attack lasers discussed in Chapter 3, the weapon's designers had to understand how the beams would be distorted by atmospheric processes. Radars must be able to lock on to targets and track them precisely, so researchers had to understand how atmospheric conditions might interfere with radio waves.

When civilian and military interests overlap in this way, we see a melding of research activities: civilian researchers watch military explorations and the military retains a keen interest in civilian discoveries. But it is not always so easy to separate the two communities in this way. There is a grey area between military and civilian space activities, involving military funding of university research and international space projects that engage participation and funding from friendly nations. Even if these are totally civilian-oriented programmes, they provide a facade under which military activities can gain legitimacy and cooperation.

CIVILIAN SPACE PROGRAMMES

Civilian space activities mostly fall under the category 'remote sensing', a relatively new term which was introduced to the vocabulary in the late 1960s. It refers to high-resolution photography of the Earth's surface and measurements of the pressure and temperature of its atmosphere, enabling meteorologists to predict the weather. In 1973 the Earth Resources Technology Satellite (ERTS 1) opened up new fields of application for remote sensing, helping us to improve our knowledge of the Earth and how human activities affect it. By 1978 this programme was called High Energy Astronomy Observatories (HEAO). Under this programme, in 1978 Einstein's Observatory, and more recently the Hubble Telescope, were placed into orbit.

By the end of 1986 environmental satellites were being operated by the European Space Agency (ESA) and by national programmes in India, Japan, the Soviet Union and the United States. The World Meteorological Organization (WMO) established a Global Atmospheric Research Program, which subsequently set up the World Weather Watch (WWW). A series of satellites in geostationary orbit now provides information on cloud motion, sea surface temperatures, analysis of clouds, and upper tropospheric humidity. This data is widely used in weather forecasting, identifying severe storms or atmospheric fronts, in disaster warnings, monitoring of shipping routes, and in aircraft navigation. Satellite imagery contributes to ice reconnaissance over the oceans, seas, large lakes and other large bodies of water. It is also a vital source of information on activities that threaten the Earth's natural balance, such as the burning of fossil fuel and the destruction of the rain forests. It is hoped that in the future, satellite sensors will provide complementary data on the space-time distribution of total ozone concentration and on the sun's radiant energy contributions to the Earth's atmosphere. This will be invaluable to researchers monitoring the so-called 'greenhouse' effect.

In the US, data for 80 per cent of the globe is collected by polar orbiting satellites. These measure the atmospheric temperature and humidity, surface temperature, cloud cover, snow cover, water-ice boundaries, ozone, and proton-electron flux near the Earth. Their search-and-rescue capability means they can pinpoint balloons, buoys, ships and remote automatic stations around the globe. They are able to monitor and predict solar events and polar auroras.[1]

Although Russia and the US are still considered to be the leaders in space, both in civilian and military terms, many other nations are beginning to participate, either alone or in partnership with others. Poland, the former East Germany, Hungary, Vietnam, Cuba, Mongolia, Romania, Bulgaria, Syria and Afghanistan have all taken part in Russia's 'guest Cosmonaut' programme. Argentina and Brazil are developing small sounding rockets and have proposed a domestic communications satellite system. Mexico is also interested and is already a major user of communication satellites. Sweden is developing its own sounding rockets and satellites for polar ionospheric research, and Czechoslovakia has done significant work in space instrumentation. Israel has its own space programme, Saudi Arabia is a leading member of the ARABSAT communications satellite organisation, and Australia possesses its own tracking facilities and satellites.[2]

Fourteen European countries belong to the ESA, which has developed the Ariane family of rockets, the Spacelab Module and the Giotto probe to Halley's Comet. In 1968 Japan formed its National Space Development Agency (NASDA), which is operated by the University of Tokyo. It is recognised as an autonomous agency under the Ministry of Education. China became the fifth nation to launch its own satellite in 1970. By 1974 it had developed a cryogenic hydrogen-fuelled third stage for its rockets, an accomplishment previously achieved only by the US and West Germany. India has an active space programme which concentrates on aerospace technology and on direct economic applications. Canada has done pioneering work on communication satellites, ionospheric satellites for research on plasma physics, and robotics and control systems, including the now famous Canada Arm (CANADARM) that can manipulate or retrieve objects in space.

This global interest is important in its own right, but is also, perhaps unconsciously, preparing the personnel and hardware for the military forces of the future. I am reminded of the 'peaceful atom' programme stemming from President Eisenhower's address to the United Nations in 1954. The UN responded to this address by creating the International Atomic Energy Agency to promote the peaceful use of nuclear energy throughout the world. This in turn encouraged universities to teach nuclear physics and engineering, and promoted a more tolerant attitude towards nuclear waste and transportation. Everyone could work for the peaceful atom in good conscience, believing that their work bore no relation to nuclear weapons technology.[3]

Of course the use of satellites to study climate change and the effects of human activity is an example of the positive potential of research, addressing the planet's most urgent problems and attempting to rectify the damage. The problem is that because leaders are still 'addicted' to war, we cannot guarantee that any research will be free from military exploitation.

MILITARY EXPLOITATION OF CIVILIAN GEOPHYSICAL RESEARCH

Because of the intimate connection between Earth's atmosphere and its weather, it is not surprising to find that military activities have had an impact on local and regional weather patterns. Back in 1946 General Electric Corporation, one of the US military's prime contractors, discovered that dropping dry ice in a cold chamber created ice crystals identical to those found in clouds. Within months this led to experiments in which dry ice was dropped from planes into cumulus clouds, converting the water droplets in the cloud into ice crystals which fell like snow flakes. By 1950, researchers found that silver iodide had a similar effect. This was the impulse behind the weather modification national research programme launched by President Eisenhower in the late 1950s.[4]

The US military's intention to undertake environmental engineering, especially to gain control of weather, is well documented.[5] Apparently the Department of Defense experimented with lightning and hurricanes in Project Skyfire and Project Stormfury during the Vietnam War. Zbigniew Brezinski, who founded the Institute on Communist Affairs at Columbia University and was an advisor on foreign affairs to Presidents John F. Kennedy and Lyndon Johnson, discussed ways of using an electronic beam to ionise or de-ionise the atmosphere over a given area.[6] According to Lowell Ponte, author of *The Cooling*, the military also investigated whether lasers and chemicals could damage the ozone layer over an enemy, causing damage to crops and human health through exposure to the sun's ultraviolet rays.[7] Canada was a partner in this research right from its inception. As early as 1962, it launched satellites into the ionosphere and began stimulating the plasma, apparently just to see what would happen.[8]

The United Nations General Assembly became alarmed at this manipulation of the weather and on 10 December 1976 the UN General Assembly approved a Convention on the Prohibition of Military or Any Other Hostile Use of Environmental Modification Techniques. This Convention was promulgated on 27 October 1978 after vetting by the UN Legal Counsel.[9] However, by labelling projects as peaceful programmes – 'pure research', 'solar energy projects' or 'industrial resource development' – governments were able to avoid censure.

ATMOSPHERIC MODIFICATION EXPERIMENTS

Atmospheric modification experiments can be categorised as either chemical or wave related. In the former, chemicals are introduced into the atmosphere causing reactions that may or may not be visible from the Earth's surface. In wave experiments some heat or electromagnetic force is introduced which interrupts or distorts the normal wave motions of the upper atmosphere. Both types of experiments have been undertaken over the past 40 years.

Red and Blue Clouds in the Sky

On 25 July 1990, the US military launched a satellite containing 16 large and 8 small canisters of chemicals – primarily barium and lithium.[10] These were released at timed intervals at a height of 32 kilometres, just above the ozone layer. This activity was repeated on a larger scale, and at different heights, in January 1991, when the US air force paid $170 million and NASA added another $81 million to create a spectacular light show over North America. It could be seen as far away as parts of Western Europe and South America.[11] The chemicals dumped into the stratosphere from canisters were simultaneously augmented by rockets launched from Puerto Rico, in the Caribbean, and the Marshall Islands in the Pacific Ocean. The sun's rays ionised the chemicals, producing luminous clouds that began as a pinpoint of intense red and blue light and spread to one-third the size of a full moon in about 30 seconds.

On 10 November 1991, an aurora borealis appeared over several American states and was even visible from Texas, something that had not happened there in recorded history. The Associated Press release described it as: 'One of the most spectacular displays of Northern Lights in years [which] awed sky gazers from Ohio to Utah and as far south as Texas, where solar particles fuelled ripples, curtains and clouds of night brightness.' Julia Penn, who lived near Chicago, described it graphically: 'It was Christmas colors! My kids were yelling Santa Claus is coming! Santa Claus is coming!' Some people called the emergency telephone number, 911, because they thought the red glow was a fire. Several fire departments responded, thinking the same thing. John A. Simpson, a University of Chicago physics professor, speculated that a sun flare had hit the atmosphere, causing the air molecules to glow.[12] The November light show was apparently not deliberate, but, of course, we cannot be sure. Whether it was deliberate or not, most scientists, when pressed, admitted that the ionosphere must have been weakened at the time. Instead of being captured in the upper layers of the atmosphere, electrically charged particles were hitting the Earth's lower atmosphere. No one seemed willing to speculate on why the ionosphere had been weakened, or why the phenomenon was occurring further south than ever before. It would not seem unreasonable to link it to the fact that just prior to this display, the military had been dumping toxic chemicals into the ionosphere.

The US and Canada had been cooperating in weather modification experiments since 1958. Black Brant rockets, made in Winnipeg, Manitoba, had been used for many years to propel Chemical Release Modules (CRMs) into the upper atmosphere. In February 1983, CRM releases into the ionosphere caused an aurora borealis over Churchill, Manitoba. In March 1989, two Black Brant Xs and two Nike Orion rockets were launched, releasing barium at high altitudes and creating glowing artificial clouds which were observed from as far away as Los Alamos, New Mexico, the US nuclear weapons laboratory.

The Churchill CRM programme involved various barium compounds, including barium azide, barium chlorate, barium nitrate, barium perchlorate and barium peroxide. All are combustible and most are destructive of the ozone layer. In a 1980 programme, some 2000 kilograms (4400 pounds) of chemicals were dumped into the atmosphere, including 1000 kg (2200 lb) of barium and 100 kg (220 lb) of lithium.[13] Lithium is a highly reactive toxic chemical that is very easily ionised by the sun's rays. This increases the density of electrons in the lower ionosphere and creates free radicals that are highly reactive and capable of causing further chemical changes.

The amount of energy in the charged particles of the ionosphere is enormous in 'Earth terms'. The most energetic particles produced on Earth are those emitted from radioactive materials, especially those created in nuclear reactions, most of which have energy below 1.5 MeV (million electron volts). The nucleus of all atoms is composed of neutrons and protons, with electrons in orbit around this centre. Protons have a positive charge but because the number of protons in a nucleus equals the number of electrons orbiting it, atoms are electrically neutral. Of course, this changes when the sun's rays energise the atoms, giving an electron an escape velocity. The energy of protons originating from intergalactic space ranges from 100 MeV up to astronomically high amounts. These make up about 10 per cent of the total charged particles in Earth's upper ionosphere. Protons that originate from our sun make up the rest of the large charged particles, and their energy ranges from 1 to 20 MeV, which is still very high in Earth terms. These high-energy particles are affected by Earth's magnetic field and by geomagnetic latitude (distance above or below the geomagnetic equator). The flux density of low-energy protons at the top of the atmosphere is normally greater at the poles than at the equator but it also varies according to sun activity. Changes in the ionosphere bring about corresponding changes in Earth's weather and climate.

Chemical experimentation with the Earth's atmosphere was undoubtedly linked to the military's desire to tap into this immense source energy and to control weather. Reports on the environmental impact of these experiments are non-existent since they predate the legislation that would have required it. I once approached the Canadian Parliamentary Librarian to see if there were official accounts of the aftermath of these experiments. I was told that there were no environmental problems since the scientists conducting the experiments did not report any and there was no outcry from the public. Of course, the public had no idea that the beautifully coloured sky they observed may have been caused by deliberate experimentation.

Probing the Atmosphere

The second type of experimentation involved the use of 'waves'. A wave is a 'travelling disturbance that transports energy but not matter'.[14] We are probably most familiar with mechanical waves, which involve the disturbance of a medium such as air, water or land – the waves in the sea, for example. The water moves up and down as the wave passes, but it is the wave energy that we see moving towards the shore, not the water itself. If there is a buoy in the water it is disturbed by the wave, but it doesn't move towards the shore. Sound waves are similar: they move through the air, water or land and carry energy and information but do not transport matter itself. We use ultrasound waves to probe the human body, and to form images of a foetus in the womb. Earthquakes send mechanical waves through the Earth that can be used to image underground structures such as oil deposits.

When water or air moves or is disturbed in a direction at right angles to the motion of the wave, we describe it as a transverse wave. There are also longitudinal waves, where the movement of matter is back and forth in the same direction as the energy flow. The movement of this wave can be compared to that of a spring coiled inside a long pipe and anchored at both ends. If you compress the spring at one end, it expands the next section. When you let go, it goes back to its original shape, compressing the next section which in turn expands the next section, and so on. This sets up a wave where the local movement of matter is back and forth, but the wave itself seems to move from one end of the spring to the other. Only a compressible substance can 'carry' a longitudinal wave. Of course, while you cannot compress and stretch a liquid in the same way you can a coiled spring, it is true that all matter can to some extent be compressed. Sound waves are longitudinal waves and can therefore be propagated in air, water or land. A completely contained body of water (for example, water in a filled, stoppered bottle) cannot carry a transverse wave but it can carry a longitudinal wave. We will see how this simple fact is used in geophysical studies of Earth.

In a complete wave cycle, energy is displaced in one direction, then rebounds, returning to its starting point. The frequency of the wave is the number of cycles that pass a given point per unit of time, usually per second, and it is normally measured in Hertz. The human ear is able to hear sound frequencies between 20 Hertz and 20,000 Hertz. Below 20 Hertz is called infrasound. When large buildings vibrate, they emit this type of sound. Sounds above 20,000 Hertz are called ultrasound, and it is this frequency that is used for medical diagnostics, locating underwater objects, analysing materials and microscopy.

Small ripples on the surface of a pond may travel about 20 centimetres in one second, while sound travels through the air much faster, at 340 metres per second. An earthquake moves through the Earth's crust at speeds as great as 6000 metres per second.

The amount of energy carried by the wave is called its intensity, and this is measured in watts per square metre of surface impacted. At 4 metres (12 feet) from a loud rock band, the sound intensity is about 1 watt per square metre; the sound from a jet plane at 50 metres (150 feet) is about 10 watts per square metre; the intensity of solar radiation on Earth is fairly constant and is equal to 1370 watts per square metre; the energy inside a microwave oven is about 6000 watts per square metre; and at a point 5 km (3 miles) away from the epicentre of an earthquake measuring 7.0 on the Richter scale, the wave energy is 40,000 watts per square metre.

Electromagnetic waves such as radio waves, microwaves or visible light have many of the properties of mechanical waves. They too can carry sound, for example, but they travel faster and can pass through the vacuum of space. Very low frequency electromagnetic waves also have the ability to pass through the solid Earth and the oceans. This capability has been very useful to the military – for example, in transmitting messages to submerged submarines. Normally electromagnetic energy comes to Earth daily from the sun. Humans, however, have reversed this process and deliberately used electromagnetic waves to probe both the upper atmosphere and the inner structure of the Earth.

As early as 1966 researchers at Pennsylvania State University built and operated an ionospheric heater, using electromagnetic energy to stimulate or heat the bottom of the ionosphere. Because the device caused problems for pilots, it was removed to a more remote location, Plattesville, Colorado. By 1974, similar research facilities were located at Arecibo, Puerto Rico, and in Armidale, New South Wales, Australia. Alarmed at this further atmospheric experimentation, a US Senate subcommittee attempted to bring some measure of accountability to weather modification in 1975 by requesting that all experiments be overseen by a civilian agency answerable to the US Congress. Unfortunately the bill was not passed.

In 1983, the Plattesville heater was again moved, together with its transmitter and antenna, to the rocket launch site at Poker Flats, Alaska. The operating contract was given to the electrical engineering department of Pennsylvania State University, under Dr Anthony Ferraro. The university operates the ionospheric heater for the US navy. Another research facility, operated by the Plasma Physics Laboratory of the University of California, is the High Power Auroral Stimulation (HIPAS) project, located at Two Rivers, Alaska, a site 40 km (25 miles) east of Fairbanks. Through a series of wires and a 15-metre antenna, this facility beams high-intensity signals into the upper atmosphere, generating a controlled disturbance. HIPAS occupies 48.6 hectares (120 acres) and contains an 'eight element, circular array of crossed dipoles'.[15] A crossed dipole is a set of five transmission towers, with one in the centre. The other four form a square around it, and the diagonals of this square intersect at the central transmitter. Two more towers are added to form a rectangle (three towers along each length), and a third is added at the intersection of the two new diagonals, creating an eight-element facility with two crossed dipoles.

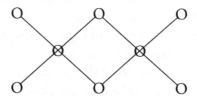

HIPAS was the first attempt to use the ionosphere to solve the navy's problem of communication with submerged submarines. In 1990 work began on an altogether more ambitious heater at Gakona, Alaska.

High-Frequency Active Auroral Research Program (HAARP)

The geographical setting of this project is halfway between Anchorage and Fairbanks, Alaska, just off one of the most beautiful drives into the foothills of the Canadian Rocky Mountains. Travelling along this highway is a lesson in the effects of global climate change, since it was built on Alaskan tundra that is starting to melt, giving the roadway a spongy feel. I noticed some of the telephone and electrical utility posts leaning at strange angles because of the shifting ground. The elegant array of mathematically positioned transmitters which make up HAARP may well become another victim of this Earth movement.

Although HAARP is called a civilian project, it is totally funded and directed by the US Air Force Research Laboratory and the US Office of Naval Research. Small grants ensure a steady stream of students from the University of Alaska and other US universities, so the programme is usually perceived as a university project.

The stated aim of HAARP is to 'understand, simulate and control ionospheric processes that might alter the performance of communication and surveillance systems'. The HAARP system will beam 3.6 megawatts (million watts) of high-frequency radio energy into the ionosphere in order to (in the words of its proposal):

- generate extremely low frequency (ELF) waves for communicating with submerged submarines,
- conduct geophysical probes to identify and characterize natural ionospheric processes so that techniques can be developed to mitigate or control them,
- generate ionospheric lenses to focus large amounts of high frequency (HF) energy, thus providing a means of triggering ionospheric processes that potentially could be exploited for Department of Defense purposes,
- induce electron acceleration for infrared (IR) and other optical emissions which could be used to control radio wave propagation properties,
- generate geomagnetic field aligned ionization to control the reflection/scattering properties of radio waves.[16]

My reason for listing the aims of this civilian-military project is not to 'lose' or mystify the reader, but to show how disconnected these goals are from normal university research aspirations. It is not necessary to understand all of the science to realise that HAARP will be modifying the ionosphere, a highly complex life support system, in order to aid military objectives.

HAARP was chosen by a prestigious panel of journalists for publication in the magazine *Project Censored*, as one of the top ten under-reported news stories of 1994. One of the pioneers in digging up information about the project is Nick Begich, a medical doctor, a skilled naturopath, and eldest son of the late US Congressman Nick Begich. Dr Begich released the first major story on HAARP in October 1994 in *Nexus*, an international magazine published in Australia. He and Jeane Manning, an experienced science journalist who had spent a decade researching unconventional energy sources, produced the book *Angels Don't Play This HAARP*, exposing what can be gleaned about HAARP from publicly available documents.

It makes chilling reading.[17] There is also a pamphlet available from the Office of Naval Research that supplies basic information. However, it is always necessary to read military descriptions of their own experiments with caution. The US Department of Defense provides a manual for its industrial contractors which states: 'Cover stories about projects must be believable and cannot reveal any information regarding the true nature of the project.'[18]

The first stage of the HAARP project was completed in 1995, and it consisted of a 3-by-6 grid of 18 antennas, or transmission towers. These are basically synchronised ionospheric heaters. The masts of each tower are 66-feet high, and they are placed at 80-foot intervals. The facility also includes diagnostic tools to measure the effects of ionospheric heating. Experiments were scheduled to take place in September 1995 and throughout 1996. By 1998, 48 transmission towers had been erected on the site, in a 6-by-8 grid formation. These were apparently funded by the US Congress through a relatively small grant of $10 million under a project entitled 'Nuclear Counter Proliferation Efforts, for HAARP Project'. This is the first obvious indication that HAARP is part of the space shield concept. This low figure is likely to be deceptive as HAARP is designed to interface with several other well-financed projects. By 2002, the HAARP site will have 180 transmission towers (12-by-15) covering 12.4 hectares (30.6 acres).

One aim of the HAARP project is the generation of extremely low-frequency (ELF) waves. Basically the transmitters can converge their beams on the electrojet and when the synchronised rays hit it at a right angle they cause the river of electromagnetic energy to spread sideways. When the rays are turned off, the jet returns to normal. If the transmitted rays are turned on and off in a rhythm, the motion outward and inward creates an alternating current that generates pulsed ELF waves. These low-frequency waves are reflected back to Earth and can be used for communications with submerged submarines and for 'deep Earth tomography', which involves 'scanning' the internal structure of the planet and which will be discussed later in this chapter.

An electronics researcher from Albany, New York, not fettered by military restrictions, explained in plain terms what he believes the HAARP experiment will do:

HAARP will not burn holes in the ionosphere. That is a dangerous understatement of what HAARP's giant gigawatt beam will do. Earth is spinning relative to thin electric shells of the multi-layer membrane of ionospheres that absorb and shield earth's surface from intense solar radiation, including charged particle storms in solar winds erupting from the sun. Earth's axial spin means that HAARP – in a burst lasting more than a few minutes – will slice through the ionosphere like a microwave knife. This produces not a 'hole' but a long tear – an incision.[19]

Even though the ionosphere is depleted and 'repaired' naturally through the action of the sun, it is not known how the atmosphere will react to these man-made incisions. Everything in our universe is in a dynamic equilibrium and this interference may destabilise a system that has established and maintained its own cycle for millions of years. As an analogy, it is normal for humans to spend part of each day awake and part asleep. However, artificially induced sleep and/or periods of wakefulness can result in unexpected problems and significant disruption of body rhythms. If experimentation with the natural rhythms of the ionosphere is potentially harmful, what would be the consequences if HAARP were used as a weapon of war?

HAARP is based on a series of US patents obtained by Dr Bernard J. Eastlund, a respected physicist and president of a technology company in Houston, Texas. Eastlund received his degrees from Massachusetts Institute of Technology and Columbia University. His patents, which he had gained while working for ARCO,[20] relied heavily on the work of Dr Nikola Tesla. In his patent Dr Eastlund described potential uses of a system of ionospheric heaters:

large source of energy, such as available in a large oil or gas field, or from nuclear power, could be used to produce electricity which would then be used to generate electromagnetic waves in the RF [radio frequency] region (from 1.5 to 7 MHz in this study). The RF waves were then to be focused by a large phased array antenna at points in the ionosphere at altitudes of 150 km and above to produce local field strengths high enough to accelerate electrons to relativistic energies.[21]

The term 'relativistic energies' means that these electrons will be travelling at close to the speed of light. Such energetic electrons are characteristic of plasma, so this could be linked to the 'space shield' concept in which areas of the ionosphere are energised to block and destroy incoming weapons.

The Eastlund patents also describe 'altering upper-atmosphere winds – so that positive environmental effects can be achieved... For example ozone, nitrogen, and other concentrations in the atmosphere could be artificially increased'. In theory HAARP could create rain in drought-ridden areas, decrease rains during flooding and direct hurricanes, tornadoes and monsoons away from populated areas.

HAARP is not unrelated to the Star Wars scenario. Tesla, whose work contributed to the development of electromagnetic weapons, was in his 80s in 1940 when World War II was threatening to disrupt Western civilisation. He devised what he called a 'teleforce', by which 'aeroplane motors would be melted at a distance of 250 miles, so that an invisible Chinese wall of Defense would be built around the country'.[22] Tesla was not able to explore his ideas fully before his death in 1943.

According to Eastlund, the current HAARP project is not yet sufficiently sophisticated to accomplish all of the military goals imagined. But even the military communication potential alone places HAARP high on the list of projects needed for 'Star Wars'. What the facility will be capable of when it reaches the final stages of its development is, as yet, unknown. However, it is certainly worth contemplating the possibilities.

One of the prime goals of the project is manipulation of the electrojet. If the electrojet touches down on Earth, it can blow out a major power grid, thus depriving a large region of electricity. Perhaps it can also be used to 'deposit energy' (a military euphemism for causing an explosion) at some point on the Earth. When HAARP is completed, it will be able to warm specific areas of the ionosphere until they produce a curved-shape lens capable of redirecting significant amounts of electromagnetic energy. These reflected electromagnetic beams may be in the microwave or ultraviolet range and could be used as a weapon either to incinerate a forest or oil reserve or to selectively kill living things.

Even with the veil of secrecy surrounding HAARP, some military documents reveal the extraordinary military preoccupation with this technology. For example, a joint report from the Air Force Geophysics Laboratory and the Office of Naval Research states:

From a DoD point of view, however, the most exciting and challenging aspect of ionospheric enhancement is its potential to control ionospheric processes in such a way as to greatly enhance the performance of C3 systems[23] (or to deny accessibility to an adversary). This is a revolutionary concept in that, rather than accepting the limitations imposed on operational systems by the natural ionosphere, it envisions seizing control of the propagation medium and shaping it to insure that the desired system capability can be achieved.[24]

There is also evidence that the US military's aim of 'dominance' extends to ionospheric heating:

Presently, a heater in Norway, operated by the Max Planck Institute in the Federal Republic of Germany, is being reconfigured to provide 1 gigawatt of ERP (effective radiated power) at a single HF frequency. The HAARP is to ultimately have an HF heater with an ERP well above 1 gigawatt; in short, the most powerful facility in the world for conducting ionospheric modification research.[25]

In 1988 and again in 1994, Caroline Herzenberg of the US Argonne National Laboratory wrote as a private individual, rather than in her research capacity, to warn that the advanced types of ionospheric heaters being developed in the US could be used as a weapon system and might well violate the 1976 Environmental Modification Convention.[26]

Two other outspoken researchers, Dr Elizabeth Rauscher and her colleague William Van Bise, published a joint paper, 'Fundamental Excitatory Modes of the Earth and Earth-Ionosphere Resonant Cavity', which described the resonant harmonies between the Earth, Earth life forms, radiant energy from the sun and vibrations of Earth's support systems.[27] Resonance is the name given when a wave with a certain frequency coincides with the natural frequency of another system causing a greatly enhanced response. The effects of resonance are often unexpected and can be quite disproportionate to the level of input. On a grand scale, resonance is thought to have caused the divisions in the rings around the planet Saturn. Introducing various wave frequencies into the Earth system could have unknown and unexpected resonance effects.

This massive transmitter poses other problems. The Federal Environmental Impact Statement filed by the air force for HAARP says that its transmissions 'can raise the internal body temperature of nearby people, ignite road flares in the trunks of cars, detonate aerial munitions used in electronic fuses, and scramble aircraft communications, navigation and flight control systems'. Even small increases in electromagnetic radiation can cause health problems such as cataracts and leukaemia, also altering brain chemistry, blood sugar levels, blood pressure and heart rates.[28]

Instead of responding to the legitimate concerns of scientists, reporters were fed stories from some unknown source about irrational fears being circulated to alarm the public:

The rumors are buzzing across the Internet that a Pentagon physics experiment on a wind-whipped tract of US Air Force land in Alaska has a secret purpose – digging up bodies of UFO aliens. Another rumor has it that men in black suits... are jumping out of a black sedan to beat up Alaskan opponents of the project.'[29]

Ridicule is a good way to scare serious physicists away from investigating a project, since they are wary of associations that might ruin their reputation or disrupt their flow of grant money. And, with such press coverage, who would listen to these 'bizarre' charges anyway? It was a very effective tactic, and there has been no concerted effort to respond to the serious call for investigation of HAARP.

SuperDARNS

The HAARP project is linked to other research and military facilities, so it cannot be considered in isolation. In fact no military project can be understood in isolation from all other projects, since they are planned to interact. One such facility is the network of SuperDARN radars. The stated purpose of SuperDARN is that they form a 'network of high latitude HF radars which will contribute to the goals of US Arctic Research Initiatives'. These goals are 'to improve the predictability of disturbances in space and their effects on high altitude communication, electric power grids, satellite orbital stability and defense systems'. SuperDARNS are supposed 'to improve our physical understanding of the electro-dynamical and mechanical coupling among the magnetosphere, the ionosphere and the atmosphere'.[30]

This is interesting because the official web site for HAARP states 'the downward coupling from the ionosphere to the stratosphere is extremely weak and no association between natural ionospheric variability and surface weather has been found',[31] yet it is clear that SuperDARNs are intended to work with the HAARP facility. To quote from the SuperDARN grant proposal: 'We also intend to begin immediately, collaborative superDARN experiments with the HAARP and HIPAS facilities (these are referred to as 'ionospheric modification facilities') and to begin collaborations with experiments operating from the Poker Flats Research Range'. In another proposal, the SuperDARN is called '...diagnostic of ionospheric modification', which modification will be undertaken by HAARP.[32] When HAARP or any of the other ionospheric heaters causes a change in the ionosphere, or when an external factor such as sun flares affects Earth's outer protection, SuperDARNs will monitor what happens in the lower atmosphere.

Each SuperDARN has 16 transmitters, a receiver, a phasing matrix and a computer. They run continuously for 24 hours, 365 days a year, and are operated remotely from Johns Hopkins University in Maryland.[33] Canadian SuperDARN facilities are located at Goose Bay, Labrador; Kapuskasing, Ontario; and Saskatoon, Saskatchewan. They were built without much discussion, and the Canadian public are generally unaware of their existence. In 1999 two more SuperDARN were under construction, one in Prince George, British Columbia, and one in Kodiak, Alaska. Other SuperDARN facilities are located at Stokkseyri and Pykloybaer in Iceland; Hankasalmi in Finland; and Halley, SANAE and Syowa in Antartica. As yet, there has been no open discussion about their purpose or goals and any information on the facilities must be gleaned by sifting through university research papers and grant proposals.

Another piece in the ionospheric puzzle seems to be the SMES (Superconducting Magnetic Energy Storage) facility being built near Anchorage, Alaska. It is a ground-based energy-storage facility for directed-energy weapons. It is expected to have a magnetic field of 30,000 to 40,000 Gaus (the Earth's magnetic field is about 0.5 Gaus). Since the full realisation of the potential of the interaction of HAARP, HIPAS and the Poker Flats facility will require massive inputs of electrical energy inputs, this storage facility may be part of the plan.

Whilst some of these projects do have direct military applications, others are more likely legitimate research projects. However, they are characterised by a rashness, an uncontrolled curiosity and a willingness to experiment with our life support system – to see what it does and how it works. The risk of potentially irreversible costs to humans and to the biosphere cannot be ruled out.

USING WAVES TO PROBE THE INTERIOR OF THE EARTH

Earth Probing Tomography

To complete the military investigation of the whole Earth system it was necessary to probe the solid Earth itself, and this again involves the use of wave technology.

Knowing that enclosed liquids cannot support transverse waves, which geologists call S-waves or seismic waves such as those released in an earthquake, while solids can, allows geologists to study the interior of the Earth. S-waves leave a shadow where water aquifers, oil deposits and gas deposits are found, since these liquids will not propagate the waves. The S-waves from a major earthquake leave a shadow zone on the opposite side of the planet to the earthquake's epicentre as the seismic waves do not penetrate Earth's liquid core. Small-scale dynamite blasts and special mechanical 'thumpers' send S-waves into the Earth at lesser depths, allowing geologists to measure changes in rock density and discover the locations of gas or oil deposits.

Underground nuclear explosions send energy waves in all directions, creating both S- and P-waves (P-wave is the geological term for longitudinal waves). By studying the P-waves, which can penetrate the Earth's liquid core, scientists have discovered that there is another solid core inside the molten centre. Timing of the echoes of these waves allows for quite precise calculations of the size of this solid inner core.

Both the inner and outer cores of the Earth are very rich in iron. Physicists do not fully understand the process, but because of the planet's rotation and the convective motions (those caused by differences in heat) of the magma, electrical currents are produced in the liquid core and these create the magnetic field that surrounds Earth. We also know now that for some reason, this magnetic field has reversed periodically, approximately every million years, and the reversals have lasted for about a thousand years. Earth's magnetic field interacts with the Van Allen belts, and during magnetic field reversals, Earth's protective outer layer is significantly reduced. This means that surface exposure to particulate radiation from the sun and the cosmos is significantly increased. Some scientists speculate that evolutionary changes in species, due to radiation-induced mutations, may have accelerated during these times.

The sun's magnetic field also reverses every 11 years, giving rise to the solar activity cycle and sunspots. The complex interaction of magnetic fields in the solar system is poorly understood, yet we can assume that a rich variety of phenomena are related to these dynamic changes. Since we do not know what triggers these profound changes, experimentation with the Earth's magnetic field could have unforeseen consequences.

Military Applications of This Technology

Although there was a partial test ban on nuclear explosions in the atmosphere in 1963, that did not stop researchers taking their explosions underground. The Soviets, for example, conducted a series of underground nuclear tests carefully spaced along the Ural Mountain chain. They successfully mapped all the underground structures on their geographical territory, including oil and gas deposits. The American military, somewhat fettered by US environmental legislation, were amazed that this could be done without strong civilian opposition.

But there is also a connection between experiments above the Earth's surface and those below its surface. Ionospheric heaters such as HAARP create extremely low frequency (ELF) waves which are reflected back to the Earth by the ionosphere. These rays can be directed through the Earth in a method called deep earth tomography. Since the beamed radiation used to convert the direct electrical current of the electrojet into alternating current must be pulsed, it is reasonable to assume that the ELF radiation it generates will also be pulsed. Pulsed ELF waves can be used to convey mechanical effects, vibrations, at great distances through the Earth. By studying their 'shadow' – that is, where the vibrations are interrupted – it is possible to understand and reconstruct the dimensions of underground structures. Funding for this technology was included in the US National Defense Authorization Act for 1995. In 1996 the US Congress, while cutting the budgets for health and welfare, set aside \$15 million to develop earth-penetrating tomography.

Woodpecker

In a book by Sheila Ostrander and Lynn Schroeder,[34] it was reported that the Soviets were experimenting with the use of pulsed ELF waves, at frequencies of 10 Hertz, the kind ordinarily found in human brain activity.[35] These pulsed waves were detected worldwide. The project was nicknamed 'Woodpecker' because of the similarity between the audible sounds on radio receivers and the tapping noise of a woodpecker on a tree trunk. It is not known whether the Russian-pulsed ELF signals were deliberate or side effects of their ionospheric activities.

Nikola Tesla had theorised that electromagnetic forces could be used to induce effects such as earthquakes (see p98). The Russian Woodpecker is thought to be a contemporary form of the Tesla-Magnifying-Transmitters, which were first tested on 11 July 1935. The *New York American* headline at the time was: 'Tesla's Controlled Earthquakes', and the article explained that Tesla had succeeded in transmitting mechanical vibrations through the Earth. The Russian version was turned on on 4 July 1976 (the 200th anniversary of the US Declaration of Independence).[36]

These early experiments with ELF waves resulted in a surprising amount of cooperation between the Soviet Union and the US. According to the *New York Times* on 21 June 1977, the US shipped a 40-ton magnet (thought to be the largest in the world at that time) to the Soviet Union. A team of American scientists accompanied the magnet. It was said to be able to generate a magnetic field 250,000 times greater than that of Earth itself and was designed to be part of a more efficient magneto-hydro-dynamic generator, used to increase the power of the Soviet Woodpecker transmitters.

The 10-Hertz ELF wave can easily pass through people, and there is concern that since it corresponds to brain wave frequency it can disrupt human thought.[37] Places where ELF waves are generated, such as the submarine communication centre for the Pacific Region in Hawaii, are now restricted areas so that no humans are accidentally exposed. However, such waves may also have a profound effect on migration patterns of fish and wild animals as they rely on undisturbed energy fields to find their way. Moreover, the wider effects of deep-earth tomography are unknown. Certainly, it has the capability to cause disturbance of volcanoes and tectonic plates, which in turn have an effect on the weather. Earthquakes, for example, are known to interact with the ionosphere. 'The key to geophysical warfare is the identification of environmental instabilities to which the addition of a small amount of energy would release vastly greater amounts of energy.'[38] This reminds us of the deep 'connectedness' of the Earth environment.

On 28 July 1976 the Tangshan earthquake in China, which left 650,000 people dead, was preceded by an airglow said to have been caused by the Soviet ionospheric heater.[39] On 23 September 1977 the *Washington Post* reported a strange star-like ball of light sighted over Petrozavodsk. A similar airglow effect was reported over the American midwest on 23 September 1993, at a time of disastrous flooding. At the same time, a lightning flash, rising from the tops of the clouds up into the atmosphere, was reported. This was recognised as a new geophysical phenomenon – normal lightning flows between two clouds, or from clouds to Earth.[40]

On 12 September 1989, magnetometers at Corralitos (near

Monterey Bay, in California) detected unusual ultra-low frequency waves between 0.01 Hertz and 10 Hertz. This is the lowest range of ELF waves. These waves grew to about 30 times their first intensity, and finally subsided on 5 October 1989. On 17 October they suddenly appeared again at 2:00pm local time, with signals so strong that they went off the scale. Three hours later the San Francisco earthquake took place. On 29 March 1992, the *Washington Times* reported that 'satellites and ground sensors detected mysterious radiowaves or related electrical and magnetic activity before major earthquakes in Southern California during 1986–7, Armenia in 1988, and Japan and Northern California in 1989'. The 17 January 1994 earthquake in Los Angeles was also preceded by unusual radio waves and two sonic booms. These strange 'coincidences' have never been completely explained.

Some recent earthquakes have been significantly different from the so-called 'typical' earthquake. Normally earthquakes occur at about 20 to 25 km below sea level.[41] However, the devastating earthquake in Bolivia on 8 June 1994 took place 600 km below the surface.

While earthquakes have always taken place periodically on the Earth, their number has increased in recent years. Not all of the data we would like is available for each earthquake, since many regions of the world lack the sensitive equipment required, but it seems highly probable that some of these earthquakes have been a result of human activity, not natural forces. In a press briefing on 28 April 1997, United States Secretary of Defense William Cohen commented on new threats possibly held by terrorist organisations: 'Others are engaging in an eco-type terrorism whereby they can alter the climate, set off earthquakes, volcanoes, remotely through the use of electromagnetic waves.'[42] The military has a habit of accusing others of having capabilities they already hold!

EARTHQUAKES OF MAGNITUDE 6 OR GREATER[43]

Magnitude (Richter Scale)	1900–1949	Average/Yr	1950–1988	Average/Yr
6.0 to <6.5	1164	24	2844	72
6.5 to <7.0	1110	22	1465	37
7.0 to <8.0	1044	21	669	17
8.0 or above	101	2	30	1
TOTALS	3419	68.4	4963	127.3

Note: Some of these earthquakes occurred within two or three days after underground nuclear explosions, but not all can be attributed to underground disturbances.[44]

Deep Earth probes appear to be an integral part of the military's aim to control and manipulate natural Earth processes. Whilst the potential of ELF waves to generate Earth movements, with associated freak weather, is frightening enough, it is also clear that the interaction between the earth and ionosphere that takes place during ELF generation and transmission may be capable of inducing more direct weather effects.

GWEN

Today's wars require high-powered radio transmitters not only for communication but also for jamming the enemy's transmissions and anti-aircraft defences, to hunt targets, guide weapons, and create intense levels of electromagnetic radiation capable of disabling high-tech electronic circuitry. The GWEN (Ground Wave Emergency Network) system was originally conceived as an emergency facility for communications during a nuclear war. GWEN was to operate using ELF radiation of 72–80 Hertz. This section of the electromagnetic spectrum has a very long wavelength, 4000 km (or 2500 miles), and because of this, was considered very resistant to the blackouts caused by the electromagnetic pulse of a nuclear bomb. Also, with the latest equipment, ELF reception can be attained at 400 metres (1300 feet) below the ocean surface. VLF (very low frequency) radiation, below 72 Hertz, currently penetrates the ocean surface only to 10–15 metres (30–40 feet).

In March 1987, Colonel Paul Hanson, GWEN program director for the air force, said: 'the towers will NOT help wage a nuclear war because they would be destroyed in any protracted confrontation'.[45] So why, then, was the US government planning to build 29 more units at a cost of $11 million? Some clues as to other capabilities of the system can be gleaned from its long history.

In 1968, the US navy broke the secrecy that had surrounded its development of an ELF submarine communication system and announced that it was going to build a facility in Wisconsin that would be capable of surviving a nuclear attack. This project included a huge grid of cable buried 4–6 feet underground, and taking up about 16,828 square km (6500 square miles) of land. There would have been over a hundred unstaffed transmitter capsules. This ambitious project was never completed because of civilian opposition to it. However, the navy did build an ELF test facility in the Chequamegon National Forest south of Clam Lake, Wisconsin, stringing 45 km (28 miles) of antenna cable above ground on poles. In 1977, just about a year after the Russian Woodpecker started, local inhabitants claimed that the government conducted an ELF experiment which created an enormous downburst of rain in six counties of northern Wisconsin. Phillips, a small town in the region, was totally devastated, and 350 hectares of forest were destroyed. Storm damage was estimated at $50 million.[46]

A newsletter article described the event in depth. During the massive storm:

> the antenna began transmissions at 13:00 hours by shifting from 25 Hertz to 72 or 80 [this would be correct since the ELF signal has a simple 0 – 1 code and needs only two frequencies, one to represent 0 and the other to represent 1]. The transmissions were pulsed at a rate of 16 times a second. The ELF antenna loop used the ionosphere as an outer shell of a spherical capacitor [storing more electrical potential than the surrounding Earth], with the inner conductor composed of the Earth's surface. This circuit duplicates the process that occurs during thunder and lightning storms.[47]

Geophysicists quoted in this article had conducted an analysis of the storm, based on an aerial survey. This analysis revealed as many as 25 local centres for the storm. 'Straight line winds diverged out violently from the local centers, each in their own downburst type of configuration.' It was almost as if there were 25 separate storms in action over a limited area. There was also evidence of a direct relationship between these 'centres' and the position of ELF transmitters.

Public concern over the physical and environmental effects of the electromagnetic radiation from this project forced the navy to abandon it. The navy changed the project's name from Sanquine to Seafarer (Surface ELF Antenna For Addressing Remotely-deployed Receivers), dropped the claim that the project would survive a nuclear war, and developed an Environmental Impact study for locating it in Michigan's Upper Peninsula. Michigan was to have a very large grid, 12,168 square km (4700 square miles). However, on 18 March 1977 the governor of Michigan, William G. Millikan, vetoed the project. He is reported to have said: 'The people of Michigan do not want Seafarer, nor do I.'[48]

In response, President Carter terminated Seafarer on 16 February 1978 and called for further studies. The programme was re-activated by President Reagan, and on 8 October 1981 the Pentagon released a scaled-down version with two transmitters, one in Wisconsin and one in Michigan, connected by 'secure data links', working independently, but with a lower signal strength for each. This latter plan became operational in the 1990s, and all submarines were outfitted with ELF receivers. This is the system referred to as GWEN.

The part to be played by GWEN in future warfare, since it is not secure from nuclear attack as was first claimed, is not clear at this time, but one wonders whether its storm-making ability is one reason for retaining it. The *Bulletin of the Atomic Scientist* noted that GWEN units lay directly in the middle of the high rainfall area of a huge flood that occurred in 1993. An unusual shift in the jet stream acted as a barrier to a cold front, bringing down 150 to 200 times more rain than normal.[49] On 10 July 1993, the *New York Times* reported that the flooding rains seemed to be locked over the Mississippi area for over six weeks. By August it was known that over a thousand levees had been breached or broken in the flood, homes were destroyed and crop damage was beyond rescue.

It has been speculated that an 'electronic dam' can be set up using ELF generators – a magnetic field is created which stalls or blocks a weather front, therefore causing torrential rain over an area. This is hard to confirm or deny because of the secrecy which protects this activity. But weather manipulation is certainly possible. In 1992, the *Wall Street Journal* ran an article about a Russian technical company ELATE. Igor Pirogoff, the commercial director, stated that ELATE can fine tune weather patterns over a 200-mile range. According to Pirogoff, 'The company can clear off areas of smog, divert typhoons and chase away acid rain.'[50] In 1994 the US air force defense declared that studies on weather modification had done so well they were now a permanent part of air force operations. The air force stated that details of weather control were classified and would not be made public.[51]

Of course, this secrecy means that we can only speculate as to the true extent of the military's potential for atmospheric modification. However, it is clear that incidences of 'freak' weather are on the increase.

CLIMATE CHANGE

Much abnormal weather has been blamed on El Niño, and it became a household term in 1997–98. El Niño is caused by a periodic heating of the Pacific Ocean current which flows up the west coast of South America. This heating is thought to be related to the activity of underwater volcanoes, but it is not known why these volcanoes should have become more active in recent years. The 1997–98 El Niño was, in the words of the UK Meteorological Office, 'the most extreme on record,' and even our most sophisticated climate models were not prepared for it. It seriously affected the weather all over North America and as far away as Europe. It destroyed ocean plankton and caused fish stocks off Peru to decrease dramatically in size. It provoked violent downpours and landslides; it spawned hurricanes and widespread crop failures, drought and massive forest fires.

Until 20 years ago, El Niño occurred approximately every four to seven years and lasted for about a year. Afterwards it would give way to the opposing climate, La Niña. Where El Niño brought rain, La Niña would bring drought. Normal climate patterns would prevail between these extremes. This pattern showed a sharp change in the mid-1970s, with El Niño appearing more frequently and becoming progressively more severe. This could possibly be linked to the fact that global warming has increased the amount of moisture stored in the atmosphere. As the Earth temperature rises the atmosphere is able to absorb more moisture, which means that there is significantly more energy available to drive storms and associated weather fronts. Over the United States, moisture in the air has grown by five per cent per decade since 1973. In the temperate region of the northern hemisphere, moisture has increased overall by 10 per cent since 1900.[52]

Not all violent weather can be linked to the effects of El Niño, however. Even before the dramatic weather observed in 1997–98, there were signs that normal climatic conditions had been severely disrupted. In 1996 Canada, for example, was hit with very extreme weather. On 18 July 1996 grapefruit-sized hail cracked the windshield and dented the nose of a Canadian airline jet flying a transcontinental route from Vancouver to Ottawa. During the late spring and early summer of 1996, torrential rains flooded the Saguenay river system, killing ten people in Quebec and driving more than 12,000 from their homes. The flooding covered more than 400 square miles of farm land. Earlier, in April, a pair of tornadoes smashed through south-central Ontario, exploding houses and barns, collapsing hydro lines, killing livestock and seriously injuring many people. In mid-June tornadoes with 160-km winds tore through Saskatchewan, knocking down power lines and ripping off the front door of a post office. In mid-July hail and thunderstorms trashed a trailer park outside the city of Edmonton. Another tornado touched down in Medicine Hat, Alberta; and in Sarnia, Ontario, lightning blew off the roof at the Suncor oil refinery, igniting a gasoline additive inside and causing a seven-hour blaze. Climatologist David Phillips of Environment Canada stated: 'We're getting a little weary waiting for something normal to happen!'[53]

Nor was the weather much better elsewhere on the globe. In

1996 severe flooding occurred in Nepal, eastern India, and Bangladesh. Three million Indians were left homeless, and more than a million Bangladeshis were trapped by flooding. The death toll from the floods which ravaged central and southern China was more than 1400, as some six million people tried to maintain the dikes and flood walls along the Yangtze River. Civil defence personnel in China estimated that 56,000 people were without electricity and 90,000 were without water.

In South Africa, heavy snowfall was seen in areas that had not seen snow for more than 35 years. Villagers in the mountainous kingdom of Lesotho were unable to get food because of the snow. Many suffered from hypothermia, while others died from the fumes from coal heaters they were using to keep warm. During the week ending 19 July, two strong earthquakes, together with high winds and heavy rains, rocked the French Alps. Quakes were felt in Austria, southern Italy, northeast India, Japan's Nii-Jima Island region, central Japan, Indonesia's Sulawesi Province, the Kamchatka peninsula and southern Mexico. New Zealand's Mount Ruapehu volcano sent a column of ash and steam 6000 metres into the air. An earthquake of magnitude 6.6 occurred in the island of Sulawesi, Indonesia. Tremors were reported in Kenya, Germany, the Greek Islands, western Turkey, northern Sumatra, Bali, the central Philippines, New Zealand's North Island, eastern Japan, central Chile, El Salvador and the Aleutian Islands, all within the week ending 26 July. Mount Etna, in Sicily, also erupted during this week and its molten lava and fire were visible for miles.

On 9 July 1996, as scientists from all over the world gathered in Geneva with United Nations officials to urge tightening of the target for the UN Climate Control Convention, a Reuter's news release from London stated: 'Freak weather conditions have claimed hundreds of lives and created chaos around the world in recent days as scientists warned governments that "greenhouse gases" and global warming may distort climates.'[54]

It should be noted that the scientists have now changed the 'global warming' message to one of 'global climate change'. This is because the effects are not homogeneous around this fragile planet. While some parts of Earth are now experiencing as much as a four-degree centigrade increase in average temperature, others have become colder. The polar regions should warm more rapidly than the equatorial region, and the continents more rapidly than areas of deeply circulating oceans. Both the Antarctic and the Arctic have shown consistent warming patterns. In fact a chunk of the Antarctic glacier as big as the state of Rhode Island collapsed into the South Atlantic Ocean just before the March 1995 Global Climate Change Conference in Berlin.[55] The loss of glacial ice is significant since if this large amount of water moves from the land to the ocean, water levels will rise and low-lying coastal regions will eventually be submerged. Some scientists also believe that the cooling of the oceans with global ice could trigger another ice age by changing the direction of convection currents.

Normal life-friendly temperature fluctuations of Earth allow for water to exist in all three states: gas, liquid and ice. If it is too cold, Earth has an ice age. If it is too warm, the water will evaporate and the planet will not be able to sustain life. Earth's balance keeps the temperatures not too hot and not too cold.[56] It is not only the biosphere that needs a controlled amount of heat. Scientists have discovered that Earth's outer atmosphere is shrinking at a rate of one kilometre every five years. It is thought that as more heat is trapped near the surface of Earth, less is transferred to the outer atmosphere. The effect of this change on life is unknown.

Trapped heat near Earth's surface is thought to be due to the release of 'greenhouse gases': carbon dioxide, methane, sulphur and nitrogen oxide derivatives, and chlorofluorocarbons. The chart shows how human activities have changed the concentration of these gases in the atmosphere, and which particular activities have been thought responsible. When environmental crises occur, it is usually only the civilian economy that is called upon to rectify the balance, while military programmes are rarely taken to task.

INCREASE IN GREENHOUSE GASES – PRE-INDUSTRIAL ERA TO 1990S[57]

Greenhouse Gas	Change in atmospheric concentration	Length of time it persists in the atmosphere	Human activity which causes it to accumulate
carbon dioxide (CO_2)	280 ppm to 365 ppm	around 200 years*	burning fossil fuels (coal, oil and gas)
methane	700 ppb to 1720 ppb (about 20 times more potent than CO_2)	about 12 years	deforestation, growing of rice and raising cattle, leaks from natural gas lines
nitrous oxide	275 ppb to 310 ppb (about 200 times more potent than CO_2)	about 120 years	modern agriculture and heavy application of chemical fertilisers; use of automobiles
chlorofluoro-carbons	from zero to 280 ppt for CFC 11 and 484 ppt for CFC 12 (many thousand times more potent than CO_2)	several thousand years	refrigeration and air conditioning, including that of planes and space ships

* There is a lag time between the release of CO_2 and its impact on climate of 50 to 80 years.

Key: ppm = parts per million, ppb = parts per billion, ppt = parts per trillion

The upward trend in Earth's temperature has been recorded over the last 130 years. It has been a somewhat jerky trend, apparently modified by industrially generated sulphate aerosols, which reflect the sun's rays upward and cool the Earth. These sulphate aerosols only remain in the atmosphere for two weeks at most, and the distribution of their release depends on the location of industries and sulphate production activity. During World War I and World

War II the production of sulphate went up measurably and the effects of Earth warming were counteracted. During the Great Depression sulphate emissions decreased markedly, and the effect of greenhouse gases increased. As we institute scrubbers on factory stacks to clean up industrial sulphate emissions, we will experience the full brunt of the greenhouse gases.[58]

Sulphate has other properties that have had an impact on the biosphere, however. Dr Harold Harvey, studying the effects of sulphur dioxide and nitrogen oxides on lakes in Canada, coined the term 'acid rain' to describe their acidifying effect. By interacting with water vapour in the air, sulphate forms sulphuric acid and nitrogen oxides form nitric acid. This acidification destroys much of the ecosystem. Acid rain has posed serious problems for North America, Europe, China and to a lesser extent Brazil, South Africa, Venezuela and Australia. It is usually blamed on sulphur dioxide, a by-product of industrial production and, especially, on fossil-fuelled electrical turbines. However, the input of the nitrogen oxides released during nuclear atmospheric testing[59] and from nuclear reprocessing plant has never been officially estimated.

This discussion of sulphate is not an appeal for more pollution to cool our overheated Earth, but rather a caution against increasing particulates in space, such as the disastrous copper needle experiment and the debris left orbiting the planet by other military space experiments.[60] It is estimated that there are 10,000 to 50,000 objects larger than four cm currently in Earth orbit. The total number of debris particles is much larger. They travel at a very high velocity and can do considerable damage to, or even destroy, a large satellite.[61] A holistic analysis of climate and weather would have to include these orbiting particles.

Some heating and warming cycles of Earth are natural, of course. The Earth orbit, for example, shifts from being circular to being elliptical over a period of 100,000 years. Its tilt on its axis varies from 21.8 degrees to 24.4 degrees over a 40,000-year period. The greater the tilt, the more extreme the seasons. Whether the northern or the southern hemisphere is closest to the sun during its summer or winter changes over a 25,000-year cycle. The northern hemisphere is now closest to the sun during its winter and farthest during its summer. This means that it receives about five per cent less summer sunshine than it did 12,000 years ago. According to some analysts, we are heading into another ice age since these seem to last 90,000 years with a 10,000-year interim period.

If the reader feels confused at these conflicting signals, it is because we do not know enough about the natural cycles of Earth and the impact of human activities on it to make good predictions of what will happen when human activities interfere with them.[62] Moreover, such predictions are based on the natural history of our planet and are meaningless in the face of random experimentation on major Earth systems in the upper atmosphere and the bowels of the planet.

Any natural stasis is maintained by having two opposite tendencies and a trigger for each. For example, the level of sugar in human blood is increased when it is low by the action of adrenalin on the liver which releases stored sugar, and when it is too high it is reduced by the action of insulin released from the pancreas. These are physiologic triggers that recognise the blood sugar level as 'too high' or 'too low' and initiate corrective actions. It is not at all unusual to discover such opposing but complementary forces in a well-balanced Earth system. One example is the cycle of growth in temperate climates. In summer, the absorption of carbon dioxide from the atmosphere by green plants and leaves is at a maximum, providing a slight cooling of the sun's heat. In the winter, the green leaves are gone and the plants are in a dormant state, allowing carbon dioxide to accumulate and creating a slight warming effect. Obviously human activities have failed to respect this helpful cycle and have overwhelmed the mechanisms by the dual action of forest destruction and excessive carbon dioxide generation.

ACTS OF GOD?

There is a long learning curve between causing freak weather by accident and being able to cause it whenever and wherever one wants. But when we consider the precarious state of Earth's health, neither practice is acceptable. While some of the events of 1996 may have been 'acts of God', certainly the overall volume and ferocity was anything but normal. It has been estimated that between the 1960s and 1990s, major natural disaster rates have increased by a factor of 10. Uninsured disasters have increased by a factor of 7, and insured disasters by a factor of 15. This has caused widespread concern in the insurance business. Weather disasters between 1989 and 1995 included 13 whose costs exceeded $3 billion.[63] According to the Munich Reinsurance Corporation, one of the world's largest underwriters, the worldwide bill for severe weather damage from 1996 to 1998 was $180 billion.[64]

The Ecologist's Declaration on Climate Change is unequivocal:

Our health and food supplies will be affected dramatically by increased droughts, heat waves and the spread of disease-bearing insects and pests in response to rising temperatures. Agricultural lands and our towns and cities will also suffer substantial damage from rising sea levels, and increased flooding and violent storms with huge costs for industry and ordinary people as their homes and livelihoods are destroyed . . . Global temperatures are rising at a faster rate than for 10,000 years, with the twelve hottest years in recorded history occurring since 1980.[65]

This group of some of the world's most prestigious scientists predicts that without changes in our behaviour, we could find ourselves in a situation of catastrophic, runaway climatic destabilisation.

CHAPTER 5
THE ENVIRONMENTAL CRISES SPAWNED BY WAR-MAKING

We have seen how military activities endanger our environment through the direct effects of war and through experimentation, but another facet of the problem is the military's abuse of natural, human and financial resources that are already in scarce supply. One of the greatest natural treasures on Earth is human potential. However, warfare reflects a value system that places the acquisition and protection of wealth above the preservation of life. Each time we waste our resources on militarism, more people die. According to the United Nations High Commission on Refugees, the global community will experience an 'enormous loss of human capital over the next generation due to malnutrition and stunting', which could mean the loss of as many as one billion children.[1] The Committee on Nutrition submitted a report, 'Ending Malnutrition by 2020', calling for a dramatic new women-focused approach to ending hunger. However, without equally dramatic shifts in national priorities, there is little women will be able to accomplish.

There is also the fact that military research has led to the development of both military and civilian products that have polluted the land and have resulted in the illness and deaths of many thousands of people. In the public mind, for example, there is no connection made between the soil degradation caused by pesticides and the development of military weapons, but the link is there. These are long-term problems.

In a world of linear thinking, in which the 'bottom line' is always the financial cost, there will be a continued sacrifice of what is natural to our Earth: air, water, land, wildlife, food and human welfare. Yet wealth is useless if we have destroyed the very support system on which all living creatures depend. There is simply no point in waging wars to 'protect our assets' if we have harmed the regenerative power of the Earth itself. In a rapidly developing global economy, we must therefore ask ourselves some difficult questions: What is the true extent of national and global resources for the support of life? How much do we, the global community, use, and how efficiently do we manage what we have? How much natural capital can the Earth replace in a year? How much sustainable air, land, and water will we leave for future generations? These questions are fundamental to the continuation of life on Earth.

Although there have been major world conferences to discuss the environmental crisis, very little attention has been paid to depletion of resources caused by military production or by civilian enterprises that have arisen from military research. On an individual level, however, military wastage often motivates personal decisions. A large number of British scientists, for example, publicly refused to have anything to do with US Star Wars research, although the British government of the time encouraged them to apply for American research funds. On 30 November 1995, Sir Michael Atiyah, president of the Royal Society, stated:

> To criticise our contribution to the arms trade might be deemed naive, unpatriotic, and irresponsible. On the other hand, as a scientist, I cannot by my silence condone a policy which uses scientific skills to export potential death and destruction to poorer parts of the world, where their scarce resources would be better employed on food and health.[2]

The picture is even more disturbing when one realises that the countries exporting arms are generally importing raw materials from the same nations to which they export 'death and destruction'. This chapter takes a hard look at military activities with respect to their use of resources, with reference to the abundance of these resources and other, more pressing, priorities for their use.

UNMET GLOBAL HUMAN FINANCIAL NEEDS

Often the relationship between military and civilian priorities is described as competition for money, the choice between butter and bombs. The reality is more complex than this. The costs of war are not always local or direct, and it has become possible for military activities of one country to impact on nations not even remotely involved in the violence.

A memorandum on the impact of the Gulf War on developing countries gives a good global view of the problem.[3] This war, with the unofficial aim of securing 'first world' Gulf oil access, caused for the developing world a doubling of the price of fuel, severe curtailment of public transportation, soaring costs for kerosene needed for lamps, a doubling of the cost of airlifting food to central African famine areas, loss of salary remittance for migrant workers in the Gulf, economic crises due to the sudden return of 'overseas' migrant workers, reduced demand for exports from developing countries (especially from Kuwait and Iraq), reduced tourism, loss of foreign aid money diverted to the war effort, and secondary increased costs for freight and insurance. These raised costs amounted to more than one per cent of the GDP (gross national product) of at least forty low- and middle-income countries. This is the United Nations criterion for defining a natural disaster. The cost to Yemen, classified by the UN as a least developed country, was over 10 per cent of its GDP, and in Jordan, the costs of the war were over 25 per cent of the GDP.

Article 50 of the United Nations Charter provides for compensation to members affected by Security Council decisions, yet the World Bank failed to provide sufficient help to the forty countries of Africa most affected by the war, despite the fact that it was reporting a record net income at the time. Furthermore, the United States was said to have made a profit on the war, collecting $53 billion for its war effort, $14 billion of which came from Saudi Arabia, with Kuwait paying around $22 billion.[4]

When we consider the financial implications of warfare, it is necessary to take a wide view. Even if we do focus purely on the level of 'bombs versus butter', however, the figures are astounding. According to the UN, world military expenditure peaked in 1986–87 at about US $1 trillion. Levels have since declined to about $700 billion per year, with the US budget alone estimated to be approximately $260 billion. According to the Stockholm International Peace Research Institute (SIPRI), the only nation showing a significant increase in military spending since the end of the Cold War is China. However, precise calculations of the military budgets of countries such as China and Russia are difficult because of their secrecy. We also need to bear in mind that money funnelled through civilian research programmes and universities will not be included in defence estimates.

Although imprecise, these figures do help us focus on the skewed nature of our global priorities. At the moment nearly one-third of the world's children are said to be living in poverty. The World Summit for Children was held in September 1990 at the United Nations in New York and was attended by many world leaders. Its purpose was to publicise the coming into law of the Convention on the Rights of the Child, introduced into the UN General Assembly on 20 November 1989 and ratified by 191 countries. (It is interesting to note that the US did not sign, its stated reason being that it wants to have the option of executing children who commit violent crimes.) At the summit it was estimated that $25 billion a year would be the minimum amount required to preserve the health and safety of the world's children. This was thought to be approximately the cost of:

- safe drinking water and adequate sanitation;
- reduction of maternal and child deaths, and family planning education;
- literacy programmes;
- supplementary feeding programmes and good nutrition;
- community health measures, immunisation for all children, refrigeration for vaccines, staffing for local clinics and mobile outreach units.

In contrast, the request for Pentagon funding of the US Ballistic Missile Defense project for 2001 alone was $30.2 billion.[5] The current budget for the BMD programme is being debated and may rise to $40 billion. Even this cost is likely to change as the technology evolves. It is difficult to gain an accurate view of the ultimate cost of the whole scheme as defence projects are only costed over five years, but it has been estimated to be between $500 billion and $1000 billion. This does not include the nuclear warheads of other weapons that are paid for by the Department of Energy,[6] nor does it include university funding, veteran health care and disability allowances. Twenty-five per cent of Gulf War veterans are now on disability benefit, the highest proportion of any war.

Ruth Sevard has estimated that removing $50 billion from the global military budget would be enough to:

'• clean up nuclear production plants which are significantly polluted;
• provide safe water for one third of the world's population;
• provide supplementary food for the world's 900 million undernourished people;
• supply community health care for 1000 million of the world's poorest.'[7]

The huge amount of money spent on the military has always been justified by the argument that the weapons industry produces jobs. Nothing could be further from the truth. For example, in the UK there is a government-subsidised urban regeneration programme, which costs on average £21,600 for each job created. On the other hand, the British Eurofighter programme costs about £250,000 per job. In a serious analysis undertaken by [first names] Barker, Dunne and Smith in 1991, it was estimated that if the UK cut its military expenditure in half, it could reduce unemployment by 520,000 and increase its GDP by almost two per cent.[8] SIPRI has noted that while UK military producers made $3267 million in profit between 1990 and 1992, they also reduced the number of jobs by 89,869.[9] Just think what could be achieved on a global scale if all countries were to reallocate just 20 per cent of their annual military budgets.

The US Bureau of Labor Statistics has estimated that $1 billion could be used to produce:

'• 76,000 military-related jobs; or
• 92,000 jobs in transport; or
• 100,000 construction jobs; or
• 139,000 jobs in health services; or
• 187,000 jobs in education.'[10]

Again, this raises difficult questions about our priorities.

The main difficulty with converting production from military to civilian products is that it requires serious planning so as to avoid social and economic dislocation. The move towards a more efficient civilian industry that maximises resource efficiency and minimises wastage will require the application of our best minds. Unfortunately, the brightest young university graduates are often enticed towards high-tech space research because of government-subsidised salaries and benefits. Bill Gates, who dropped out of Harvard University before graduating, escaped this pattern and went on to construct the Microsoft computer empire, devoting his considerable talent to making computers user-friendly and affordable for the general public. Inspite of recent allegations about monopolistic practices stemming from software compatibility problems, no one has ever accused him of using his research for violent or malicious ends.

NATURAL RESOURCES

Human beings need a steady supply of the basic requirements for life: quality food and clean water; housing; energy for heating or cooling the space we live in, and for transportation and manufacturing; fibres for clothing, furniture and paper products; and efficient waste-disposal systems. Intuitively we understand that the notion of sustainability means ensuring there are enough resources for everyone and that they should be shared across national boundaries. This requires responsible resource management, keeping within the means of nature, and leaving sufficient resources for future generations. Insufficient natural resources and lack of a decent and equitable standard of living generate conflicts over water, land, food and scarce minerals or oil. This is a vicious circle because each conflict further degrades what is available for our use.

Earth has a surface area of 51 billion hectares (126 billion acres), of which 71 per cent is sea and 29 per cent land. Only 8.3 billion hectares (20.5 billion acres) is productive land, while the remainder is covered by ice, is desert, or has soil unsuitable for human use. Pollution, loss of top soil, deforestation and desertification threaten to reduce the land available for production, while over-fishing and sea dumping reduce the fertility of the seas.

The uses of this 'available land' can be subdivided into: oil, gas, coal and mineral reserves; arable land; pasture; natural forest for carbon-dioxide absorption, timber, erosion prevention, climate stability, maintenance of water cycles, and biodiversity protection; human settlements, mining resources, and sea (for food, capturing solar energy, and for exchanges of gas with the atmosphere).

If we divide this 'available' land by the number of people alive today, we can assign a per capita 'ration', made up of the following components:[11]

* 0.25 hectares (0.62 acres) arable land
* 0.6 hectares (1.5 acres) of pasture
* 0.6 hectares (1.5 acres) of forest
* 0.03 hectares (0.074 acres) of built-up settlement
* 0.5 hectares (1.2 acres) of sea

TOTAL: 1.98 hectares (about 5 acres)

However, since we humans share the planet with 30 million other species, not all of this land is really available for our consumption. Life on Earth is sustained by a complex and interdependent web of creatures. Although we seldom think about it, human life is totally dependent on the algae that feed the fish that feed the birds that feed the wildlife that feeds the domestic animals that produce fertiliser and the food we put on the table. In an ecological disaster, such as a volcanic eruption or dramatic climate change, it is possible that some obscure organism essential to the integrity of the web could become extinct. It is therefore prudent to preserve as much biodiversity as possible to ensure the existence of alternative components should one component be eliminated or disrupted.

The World Commission on Environment and Development has recommended that at least 12 per cent of Earth's resources be set aside to preserve biodiversity.[12] This means that, at best, only 1.7 hectares of resources are available per person per year. Obviously, these resources are not homogeneously spread throughout the various nations on the globe, nor are they equitably distributed within national boundaries.

Miltary Use of Resources

On a global scale, the military takes up a significant amount of land, using it for bases, testing sites, toxic waste dumping, motor repair pools, and other environmentally contaminating activities. Much of the waste is not easily recycled and its polluting effect on the environment can last for thousands of years. Additionally, the military uses significant amounts of fuel, aluminium, copper, lead, nickel, and iron ore – metals in limited supply. Since 1980, the US, Japan, Russia, Germany, the UK and France have been among the top ten importers of these metals.[13] Heavy machinery such as tanks causes compaction of land, and widespread use of pesticides, developed from chemicals used in World War I, cause pollution of land and water. According to a UN study released in 1990, nearly one-sixth of the world's vegetated regions have suffered soil degradation since World War II, and 25 per cent of this degradation has not been the result of farming. A NATO Committee identified some of the environmental problems their activities have caused:

'• leaks of toxic substances during the transport of military materials;
• atmospheric pollution over coastal areas;
• air and water pollution from ship engines;
• the transportation of contaminating agents along rivers, via deltas and estuaries;
• the dumping of radioactive waste;
• noise pollution; and
• chemical accidents.'[14]

The ultimate goal of all this military activity is even more devastation – the destruction of buildings, bridges, industries and equipment produced by the 'enemy'. Thinkers marvel at the inefficiency of the automobile, which uses only 1 per cent of the energy produced by its motor in order to move (the rest is released into the air as heat and pollution). How much more inefficient the MX missile, produced to destroy as much as possible and be destroyed in the process!

Companies, including military industries, are brilliant at externalising their costs to the environment. They take the free resources nature provides but do not factor the cost of replenishing these resources into their production process. A company may use water for cooling, for example, and this water may become unfit for other purposes, but the company is unlikely to pay a contribution towards water-purification schemes. Also, when industry pollutes the air and water, human health costs are generally charged to the individual or national health programmes, not the industry responsible. The professional journal *Nature* puts the global costs of such 'free' services at $33 trillion per year.[15] Because resources are perceived as 'free' they are wasted.

Bearing all of this in mind, one would have thought that the military's impact on the environment would have been a major topic at the United Nations Conference on Environment and Development in Rio in 1992. However, this issue was ruled out of the agenda, apparently under pressure from the United States. In the official documents of this UN conference, the US delegation had circled every mention of the 'military', disputing these sections until each reference was withdrawn.[16] The document on women's issues was the only one which successfully managed to mention military devastation, and then only in terms of its impact on women. Similarly, all mention of nuclear technology was barred, with the one exception being transportation rules for exporting nuclear waste from developed to developing countries. By inviting Hans Blix, chair of the International Atomic Energy A   g   e   n   c   y   , to deliver the keynote address, the impression was given that nuclear power is a 'solution' rather than a problem. (The IAEA is mandated by the UN to promote all peaceful uses of nuclear energy.)

It was very different at the NGO (non-governmental organisation) parallel conference in Rio, where military and nuclear issues were front and centre.[17] The wisdom of survival rests with the people, and the rising torrent of dissent visible in Rio rose to outright rejection of the economic priorities of the dominant nations at the Seattle meeting of the World Trade Organization in 1999.

Rio +5 Conference, March 1997

Humans normally concentrate on daily problems and the local

scene and agenda, with only the occasional 'long-range plan'. It is important when dealing with environmental problems, which have immense time and geographical dimensions, to create some measurements against which to gauge one's progress or losses over time. In preparation for the Rio +5 Conference in March 1997, 'ecological footprints' were developed to assess the progress nations had made since 1992. The 'Footprint of Nations' report, explaining both the methodology used and the findings, was first released at the conference.[18] This report examined the resource management of the 'big' players in the global economy – the 52 large nations that contain 80 per cent of the world's population and generate 95 per cent of the world's domestic products.

Ecological footprints were derived by looking at the biological and physical resources available to each country and then comparing these to the nation's average consumption. Available resources were taken to mean those that can be replenished by nature in a sustainable way. National resource excess or deficit therefore depends on a nation's average consumption, total national ecological resources available, and population size. In calculating these 'footprints' a spreadsheet composed of a hundred lines and twelve columns for each country was developed. The exports of a country were subtracted from and its imports added to domestic consumption.[19]

The good news is that 7 of the 52 nations are not exceeding their national ecological resources although they may be exceeding their equitable global ration. Those countries with surplus resources are New Zealand, Finland, Sweden, Ireland, Australia, Canada and Chile. Another seven countries are living very near to the limit of their national natural resources: Bangladesh, Brazil, China, Columbia, Ethiopia, India and Pakistan.

The bad news is that all other major countries, many of which have extensive weapons programmes, are running up yearly ecological deficits, with their resource consumption exceeding their national ecological capacity. One devastating finding is that in 1992, humanity as a whole was consuming over 25 per cent more resources per year than the Earth could replace; in 1997, that figure was 33 per cent. In spite of 25 years of talking about the environmental crisis, major programmes to save the biosphere, world conferences and treaties, overconsumption has not ceased. This annual global resource deficit is infinitely more serious than the global financial deficit, and more devastating to future generations.[20]

Approximately 422 million hectares of ecological resources are used each year for weapons production in the United States, Russia, China, United Kingdom, France, Germany and Japan. Based on the global average, this could provide sustainable life support for about 250 million people. Japan has the largest national resource deficit, primarily because of its large population and scarce national resources. The US consumes more per capita than Japan, 8.4 hectares (20.8 acres) per capita, but because it has much richer natural resources available, it came second. On the global level, the United States is consuming 180 million hectares (445 million acres) per year above its fair allotment. By setting free for civilian purposes the natural resources used by the world's military, even this huge US deficit could be alleviated.

Resource consumption is not necessarily linked to the standard of living. Japan, the UK, France and Germany are all considered to have a very high standard of living yet they manage to use fewer resources than the US. The per capita consumption in Japan is 6.3 hectares, in the UK it is 4.6, in France 5.7 hectares, and in Germany 4.6 hectares.[21] One of the differences is that these countries are more compact, therefore their transportation needs are lower. As electronic communications are improved, there should be less need for travel, thereby cutting down on emissions from cars and aeroplanes and on consumption of paper, easing the depletion of forests.

Nations carrying the largest national deficits are also those that import metals such as aluminium, copper, lead, nickel, zinc and iron ore and carry out large industrial manufacturing programmes: the US, Japan, Russia, Germany, UK, France and Indonesia.[22] With the exception of Indonesia, these same countries are the prime emitters of carbon dioxide, which is one of the main causes of the greenhouse effect. China and the Indian subcontinent are also countries with high emissions of carbon dioxide.[23] The ecological footprint is based on balancing how much is consumed and how much can be replenished by nature, but it does not take into account the toxicity of each item. So, for example, a footprint methodology would look at uranium acquired (mined or imported) and 'used' but not take into account the resources destroyed by its use. In the same way, the destruction of the environment caused by high carbon dioxide emissions will not show up in the footprint methodology.

The world average consumption for 1997 was 2.3 hectares per capita, a deficit of 0.6 per person. Military production alone can account for about 13 per cent of this ecological deficit. If the military hardware currently being developed is ever used in war, the reduction in available resources could be catastrophic. If we think of the long-term consequences of the Gulf War, a comparatively short-lived and localised conflict, what would be the results of a full-scale battle between two larger, more equally matched nations?

The future well-being of the global community depends largely on the health and ingenuity of the next generation; they will have to resolve the problems caused by this generation's activities. In the past, we have only assessed the cost of large-scale projects in financial terms; in the future, resource costs must be a central concern. This requires a change in thinking. If we could apply this, for example, to the US Solar Powered Satellite project discussed earlier, we would look at the resources 'cost' of 60 rectennas on Earth, 60 photovoltaic arrays in space and 60 space platforms the size of Manhattan. These would have depleted the national supply of sapphire, silver, gallium and arsenic, not to mention the heroic amounts of aluminium or steel required. The deployment of solar cells in space, rather than on Earth, makes their material components irretrievable and therefore non-recyclable. All satellites use large amounts of scarce resources such as platinum and molybdenum.

Large electrical generators needed for projects such as ionospheric heaters require a network of transmission lines which in turn consume enormous amounts of copper or fibre optics and take up swaths of land through existing agricultural fields, forests, and parks. The financial saving of using power from large generators may not be worth the loss in resources, and small local generators may be the more ecologically friendly option. Obviously air quality gains/losses must also be factored into these complicated environmental assessments.

Resource productivity and changed behaviour

We understand the principle of productivity when it comes to workers: one worker can be more productive and more efficient than another, perhaps by being faster, more accurate, needing less break time. However, we can also apply the same concept to resources: metal is more productive when it can be recycled into new products after its first, second, or third uses are terminated. It doesn't have to be dumped in a landfill! Uranium would be at the 'worst' end of the productivity scale because once used in a nuclear reactor it becomes 'high-level nuclear waste' which must be kept away from the biosphere forever.

There is also a great deal of wastage in our production methods. A German scientist claims, for example, that we could maintain the same lifestyle we have today using only one-quarter the amount of natural resources by rethinking our production methods.[24] My father, who was president of the Standard Mirror Company, discovered a way to reduce his company's need for silver for mirror backs by around 68 per cent. When painting the backs of the mirror under the old method, 32 per cent of the silver was left on the surrounding walls rather than on a mirror because of splashing and carelessness.

We need to demand both efficiency in the use of and productivity of additional resources. Industry must produce items that can be reused and recycled, and at the same time reduce resource use by better methods of production. Of course this concept is not popular with companies that want to create demand by creating products that are either disposable or have short lives.

In an equitable world, the major consuming nations need to reduce their consumption of ecological resources by at least a factor of four, or alternatively, ecological resource productivity should be increased by a factor of four, to maintain an adequate standard of living. Obviously the sane path lies in some combination of increased resource efficiency, increased resource productivity, and reduced consumption. This picture is further complicated by pollution, which reduces the usable air, land, food and ocean resources. Pollution threatens the lives of the myriad plants and animals with whom we share the planet and which form essential parts of the web of life.

MILITARY POLLUTION OF THE ENVIRONMENT

The use of depleted uranium (DU) weapons in both Kosovo and the Gulf War has polluted large areas of land for years to come. However, military pollution doesn't just happen on the battlefield. One of the most notable examples was the Love Canal disaster in the 1970s.

A subsidiary of the Hooker Chemical Company, one of the companies that produced Agent Orange and other herbicides and insecticides for use in the Vietnam War, was located in Niagara Falls, New York, near to an abandoned canal. The company used the canal bed for disposal of their containers of toxic waste, covered them with landfill, then sold the land to the local school board for $1. This same site contained uranium waste from the production of nuclear bombs for the World War II Manhattan Project.[25]

Although there was a clause in the deeds mentioning that the land contained toxic waste, the school board was verbally assured that it could build on the property. In turn, the new school attracted young families to locate nearby. However, when the rains came and the containers rusted, the toxic soup began to come up in basements and backyards. Children playing in a nearby creek experienced chemical burns, and there was at least one person suffering from a serious disease in every home on the property. In one especially poignant story, a soldier returning from Vietnam found the same Agent Orange he thought he had left behind in the war zone in his own backyard. Families living closest to the toxic waste were the first to be evacuated. In the following year, there were ten pregnancies among families living just beyond the evacuated area. Of those ten, only one baby was born healthy and without impairments, spurring the evacuation of another 1000 families.

Many of the 2.6 million Americans who served in Vietnam reported suffering from illness on their return, and a significant number of their children were born with health problems. Twenty thousand veterans won compensation for prostate and respiratory cancers,

Hodgkin's disease and spina bifida in their children from Dow Chemical Company and Monsanto, two firms that had manufactured the defoliant.[26] Yet the government continued to deny that there was a problem and products derived from the research were widely used in agriculture, golf courses and city parks.

Medical specialists in Vietnam claim that one million Vietnamese – whether combatants, civilians or their children – were poisoned by Agent Orange. Twenty million gallons were sprayed over 10 per cent of Vietnam, reducing dense jungles and mangrove forests to barren wastelands. Many children in the polluted areas were born with learning difficulties or severe impairments. The US government has refused to accept responsibility for Agent Orange damage to Vietnam and to its own veterans, with the exception of a skin rash. The government of Vietnam hesitates to make too much of the pollution because it could harm tourism and agricultural exports.[27]

At the time of the Love Canal event, I was living nearby and working at the local cancer research centre. Looking back on the incident, I find it astonishing that no one made the connection between the local problem and the wider issue of war. The Love Canal was considered to be an industrial chemical disaster, unconnected with the struggle of Vietnam veterans.

Love Canal activists went on to found the Clearinghouse, which helped citizens all over the United States obtain information on toxic waste dumps near their homes. Following this, the United States government set up its Superfund to clean up such sites.[28] Thousands of sites were identified, most being leaky landfills or on land adjoining military bases or weapon-production facilities. However, there appears to have been an attempt to separate military pollution and civilian pollution in the public consciousness. One of the primary targets of the Superfund was Canonsberg, Pennsylvania, site of the processing plant that had prepared uranium from the Belgian Congo for the Manhattan Project. As soon as the US government discovered that all nuclear and uranium research and production sights would be on the Superfund list, they relegated these to special handling by the US Department of Energy. From then on, the Superfund list contained only chemical waste dumps.

US citizens had access to environmental legislation, clean-up operations and effective compensation; those mining the uranium in the developing world did not. The raw materials for weapons production are often extracted at a cost to the health of the workers and to the local environment, as productive land is destroyed and precious resources are exported for little financial gain. In this way, the needs of high-tech military programmes intrude into the developing world. The people of Bukit Merah, a town of 15,000 near Ipoh in Malaysia, found this out the hard way.

The Asian Rare Earth Company

On 23 November 1979, the Asian Rare Earth (ARE) Company was formed in Malaysia with the intent of producing rare earth products for the electronic industry.[29] The ARE was to be jointly owned by Mitsubishi Chemical Industries, Japan (35 per cent); Beh Minerals, Malaysia (35 per cent); Tabung Haji, Malaysia (20 per cent); and Bumiputeras, Malaysia (10 per cent).

The ARE Company used a chemical process to extract yttrium from zenotime and rare earth chlorides from monozite. Zenotime and monozite are waste minerals associated with the tin mining in Malaysia. The Malaysian partners provided the raw materials and Mitsubishi bought the end products. These products are used in electronic equipment such as computers and television screens, and in laser technology, and all were exported to the United States, Australia and Japan. This brought in $16 million annually, but because of its high overheads, the ARE never paid taxes or showed a profit, so the Malaysian economy never gained much economic benefit from the new industry.

The waste generated by the ARE company, some 2250 tons a year, contained a concentration of radioactive thorium and radium that was six times over the level internationally recognised as hazardous and requiring special handling. Such waste should be isolated from the biosphere for at least 500,000 years and it continuously releases the radioactive gases thoron and radon. However, the ARE placed this dangerous material in plastic bags and discarded it in open trenches behind the plant. Dogs broke the bags open and dispersed the contaminants over a wide area, including land on which children played.

The local people began to realise the danger of this factory when two pregnant women, who both worked as cleaners in the factory, gave birth to severely disabled children. The people's complaint prompted the ARE to begin using barrels for their waste and the Malaysian government to request an inspection by the International Atomic Energy Agency (IAEA). As previously noted the IAEA is mandated by the United Nations to promote nuclear energy and related activities. However, even the three experts they sent expressed alarm at the careless handling of waste at the ARE factory. Their report recommended:

> the present stockpile of the thorium hydroxide waste be eliminated **immediately** as the drums are not closed properly, they are not protected from rain and flood, and there is no protective shielding or exclusion zone for protection against the external radiation.[30]

Twelve other safety measures were ordered by the IAEA.

The Malaysian government decided that the thorium waste should be stored for possible use in the country's nuclear research centre, the Tun Ismail Atomic Research Centre. In 1982, they proposed a storage site be located in Parit, a small town in Perak. However, the citizens of the town organised a protest, and the government abandoned its plans. In April 1984 they chose a different site, about one kilometre from Papan, a town of 1500 inhabitants. This was near land used to grow food and was also close to freshwater reserves. Trenches for the waste were constructed at the top of a hill that drained into the fertile lands below. These trenches were not only declared by the IAEA to be unsafe for radioactive waste, but they also did not conform to normal civil engineering principles for non-hazardous wastes. In October of 1985, the Ipoh High Court placed an injunction against ARE to stop all operations. They were to build a temporary waste storage facility until a permanent one could be constructed, and to clean up the radioactive debris.

On 6 February 1987, the ARE, without seeking the court's permission, declared that it had complied with the court's ruling and resumed production. The matter came before the court again in September 1987 and continued before the court, on and off, for two years.[31] During this trial, which received international attention, it became apparent that Mitsubishi had had the same difficulty with its earlier Japanese plant, and that it had moved to Malaysia after being put out of Japan for pollution which caused illness in the local population.

The Malaysian people testified before the court that they had about 51 seriously damaged children (a much higher number than one would expect in a Malaysian population that size), an abnormally high rate of miscarriage, and four children with cancer (about 20 times higher than the normal rate). They also brought in Dr Sadeo Ichikawa, a respected radio-geneticist from Japan, and myself, to measure the radiation still emanating from the factory and assess the health damage to the people. Members of the IAEA, the International Commission on Radiological Protection and the Mitsubishi parent firm attended all sessions of this trial over the two-year period and provided testimony to support the ARE. In spite of this, the court ruled against the company and told them to clean up and to move out of Malaysia. Although we were not able to demonstrate health effects in 1987, as illnesses due to radiation exposure often take years to become apparent, we had proved that radioactivity was still being emitted from this plant and that there was no way the factory could be operated without such emissions. The court rejected the industry's claims that these were 'harmless' or that naturally occurring radiation was even greater.

This incident demonstrates the cancerous metastasis of the military's needs into civilian society. Multinational corporations apparently lose money when producing military products but they continue to do so in order to stay at the 'cutting edge' of research. They make their money by developing spin-off products for which they create a civilian market. For me, this insidious invasion into civilian life is an important aspect of the military's pollution of the Earth and the ultimate cost of war.

While it is not difficult to make a case for eliminating the death and destruction caused by the military, what of the so-called benefits of military science? Is human history not marked by military victories and corresponding advances in civilisation? Are not the wealth and high standard of living following World War I the results of securing freedom? One benefit of military science, for example, is the purification of drinking water through the use of chlorine, which has saved millions of lives. An in-depth examination of this argument is in order.

The Peaceful Chlorine Program

In the past, most military programmes were quick to find some medical application that identified their product or technology as a 'need' in society. One of the reasons I became curious about the space programme was the military's push to involve pharmaceutical companies in the manufacturing of medicines in space. I had noticed this pattern before: chloroform, an anaesthesia, was developed from chlorine gas used in World War I; the nuclear technology that followed World War II produced nuclear medicine. We are in somewhat the same position now in that the technology that produced nuclear and biological weapons has also led to irradiated and genetically manipulated foods.

Sodium chloride, the salt of the sea, has always been a natural part of the environment but chlorine, separated as a highly reactive gaseous element, did not exist naturally and was first used extensively in World War I. After the war, the US introduced a clause prohibiting the use of poisonous gas and bacteriological weapons into the Geneva Protocol, the generally accepted international 'rules of war'. By the time World War II broke out, most nations, with the exception of the US and Japan, had signed it. Japan eventually signed in 1970, and the US in 1975, but it exempted gases used in 'riot control, chemical herbicides and defoliants'. The US had developed new products and wanted to keep its options open.

The scientific community was fascinated with chlorine and wanted to develop new uses for it. They soon discovered that it could combine with carbon, one of the building blocks of life, to form various new products: methyl chloride, methylene chloride, chloroform, and carbon tetrachloride. Chloroform is no longer used as an anaesthetic, because we now know that it is oxidised inside the body to form phosgene, another highly toxic and often lethal gas. Chloroform was found to be toxic to both the liver and the kidneys. Carbon tetrachloride was used for years as a dry cleaning fluid, until it was discovered that it causes severe liver damage, liver cancer and lymphatic leukaemia. It is now banned in many countries, and severely restricted in others.

By the early 1940s chlorine chemistry had become big business, and being a chlorine chemist was a life's work. It was estimated that tens of thousands of new compounds were synthesised using the element. All were widely spread in the biosphere before being thoroughly tested for toxicity; all were artificial and unnatural to our recycling Earth.

Chlorine is now well entrenched in the pulp and paper, pharmaceutical, plastic, and pesticide industries. It was and still is used directly to purify drinking water, and it has been incorporated into consumer products, textiles, photographic film, refrigerators, aerosols, rubber and farm chemicals. It is an essential ingredient in about half of the 48 commonly used chemicals that are the most damaging to our planet.

When chlorine is used to purify drinking water or bleach paper it is ultimately dumped into a nearby river or lake as waste. These natural bodies are full of decaying organic matter. This matter combines chemically with the chlorine, forming organochlorine compounds, most of which are highly toxic. One of the most frightening effects of these organochlorines is the creation of new compounds called pseudo-oestrogen that mimic the effect of female hormones in animals (including humans). This results in birth defects in offspring, reproductive abnormalities, poor survival, and feminisation of males. Pseudo-oestrogen is implicated in human breast and prostate cancer, the dramatic increase in endometriosis, and neurological and developmental problems in children.

Another evolution of chlorine technology was the development of chlorofluorocarbons (CFCs), which have given 'wings' to chlorine, enabling it to rise more easily to the stratosphere. CFCs have contributed to atmospheric pollution and the greenhouse effect.

Products derived from chlorine are now embedded in the consumer market, although there are many alternatives that could have fulfilled the same purposes. Once they are embedded, there is strong social and economic pressure against change, even when change may be wise. One of the most insidious and deadly aspects of chlorine technology has been the development of a range of pesticides that are damaging the health of humans, animals and the environment.

A pesticide called chlordimeform was introduced into commercial production in 1966 by Schering AG (Germany) under the trademark Fundal, and by Ciba-Geigy (Switzerland) under the trademark Galecron. In 1968, the product was registered for use against the insect pests of: apples, pears, peaches, nectarines, plums, prunes, walnuts, cabbage, broccoli, cauliflower, and Brussels sprouts. In 1972 it was further registered for use against the cotton bollworm and the tobacco budworm. After these approvals, chlordimeform was sold internationally under 20 different trade names.[32]

Between 1976 and 1978, Ciba-Geigy carried out a study on the effects of the pesticide on human health. In Egypt, six child 'volunteers' aged between 10 and 18 were sprayed with the product. They were not given protective clothing or respirators and suffered diarrhoea, dizziness, head and stomach aches, and other symptoms of chlordimeform poisoning.[33] The pesticide also caused a rare malignant tumour of blood vessels in 70–80 per cent of the animals exposed to it.

In 1976, Ciba-Geigy and Schering AG voluntarily suspended the manufacturing and sale of chlordimeform and recommended that it have limited use for cotton crops. Although they did not release their studies, it is clear that all of the other crops for which this product had been recommended, including tobacco, could involve pathways to human exposure. It was not until June 1985 that a civilian action group, the Pesticide Action Network (PAN), caught up with chlordimeform and declared it hazardous to health. By that time the product was being produced globally.

By 1990, chlordimeform was banned in Australia, Cyprus, Denmark, Ecuador, New Zealand, Pakistan, the Soviet Union, Thailand and Yugoslavia. It was severely restricted in Colombia, East Germany, Guatemala and the United States. This is only one among thousands of such products, and not even the most toxic.

The Bhopal Disaster showed just how lethal pesticide production can be. In December 1984, the Union Carbide pesticide plant in Bhopal, India, released toxic methylisocyanate (MIC), plus 26 other toxic gases, into an overcrowded area of sleeping people. More than 10,000 were killed in the immediate aftermath, and more than 200,000 others died or were permanently disabled over the next 12 years. The chemicals released were all involved in the production of Sevin, a pesticide, or were by-products of the process. Union Carbide allowed dangerous practices at its Indian plant that were not permitted in the US facilities, and had chosen to disregard prior accidents and warnings from the workers at the failed plant.[34]

Although this industrial disaster gave sudden visibility to the problem of pesticides, millions of other people have died or been disabled by these products. Humans have been contaminated as they tend the land, as they handle farm products, or as they eat food. A 140-page report by the London Food Commission (LCF) found that 49 pesticides cleared for use in Britain have been linked to cancer, 31 have been associated with birth defects in animals, and 61 are alleged to cause genetic mutations. The Ministry of Agriculture admitted that it did not systematically check food sold in shops for pesticide residue. Dr Tim Lang, director of the LFC, said that the Ministry's laboratory tests were so poor they could only detect 110 residues of the 426 pesticides for which they had granted permits.[35]

People in the developing world are worse off than those in first world countries. According to the Food and Agricultural Organization, in the developing world 9000 people die every year from pesticide poisoning. In Delhi, the body fat of ordinary people is storing up to 20 parts per million (ppm) of DDT; this is higher than anywhere else in the world. The World Health Organization considers 1.2 ppm to be the maximum permissible level.[36]

Although we humans are understandably preoccupied with the health of our own species, this technology has had enormous implications for the environment. The soil has become a reservoir of pesticides, solvents, herbicides and other toxic chemicals, slowly releasing them back into the biosphere over time. Twenty years after a substance is banned it can turn up in cake mixes, cereal, grains, cotton, homes and drinking water. According to Joe Pignatello of

Connecticut Agricultural Experimentation Station, 'the only sure way to get organic contaminants out of soil is to dig up the whole field and put it through an oven at 1500 degrees F (816 degrees Centigrade), and put what comes out back in the earth'. Doing this to one cubic metre of earth would cost between $500 and $1000. In Connecticut, these poisons have also invaded the ground water, contaminating it at levels up to 200 times the state's 'safe' guidelines.[37] In spite of the increased knowledge of the consequences of this irrational behaviour, chemical farming continues.

In *Silent Spring*, Rachel Carson pointed out that the narrow speciality of researchers bent on one effect, such as killing a pest, means they do not envisage the many problems that might occur in an ecosystem exposed to widespread use of their product.[38] Commenting on her book, Robert Rudd noted that '*Silent Spring* is a biological warning, social commentary and moral reminder. Insistently [Rachel Carson] calls upon technological man to pause and take stock'.[39]

Chlorine-based pesticides, herbicides and defoliants, created by the military for use in the Vietnam War, were dumped on the commercial market with a minimum of testing or legislation controlling their dispersal. These products were considered 'harmless' until proven toxic. Has this been a benefit to society? Some would claim that purifying water alone made all of this chemistry worthwhile. But there are other, better ways of purifying drinking water, using hydrogen peroxide for example, or ultraviolet light. One can surely question the use of chlorine products both in war and for domestic purposes. They are poisons to the environment, mutagens and carcinogens. They destroy Earth's ecological resources and their use is the opposite of good resource management.

TOWARDS THE FUTURE

This search into our past has admittedly been depressing. It is sad that indiscriminate use of chemical fertilisers on the land has caused massive erosion of the once fertile top soil, and over-fishing has depleted the rich fish stocks which have sustained life for hundreds of thousands of years.

The military is not only destructive in its use and misuse of natural resources, but it is also contributing to some of the most intractable survival problems of the twenty-first century. Prudent planning requires the setting aside of some land for biodiversity, nature preserves, and for emergency situations. Our social failures, manifested by wars, unemployment, underemployment, and dire poverty, seriously reduce human resource productivity, as do illiteracy and lack of appropriate education. Our scientific research has been oriented primarily towards destructive weapons or profit, not towards solving these urgent problems. We often waste resources by producing unneeded items and then aggressively marketing them. We plan luxuries for those with money to spend and neglect the large market of people subsisting below the poverty level. Video game wars are won without any ecological penalty, giving a false sense of the benign nature of real high-tech combat. Many children are suffering from attention deficit disorders, and they are killing one another with startling frequency. These are not healthy signs.

There appear to be two paths towards global stabilisation of population and resources: the first would use force, and violence, to reduce populations and limit consumption; the second would propose reducing the felt need for population increase through fulfilling basic survival requirements, providing security from violence, and increasing resource productivity. This second path has many supporters. On 4 April 2000, the secretary general of the United Nations, Kofi Annan, implored the wealthy nations to 'make bold moves to end the worst cases of poverty'. He called for urgent aid to the 1.2 billion people who live on less than $1 a day. He laid out an agenda for the twenty-first century which would provide clean water for everyone, reduce the Aids epidemic, make literacy available to all, provide the poor with free access to goods, curb traffic in light weapons and restructure the UN Security Council. It is an ambitious agenda, one that requires increasing the productive capacity of Earth's resources, reducing resource consumption, and reversing destructive activities. Can we find ways of achieving greater eco-efficiency and eco-sufficiency, peace, and rule by law? Can we accomplish an equitable distribution of goods and services both within and between nations?

I believe that there are clear steps we can take. As in most serious illnesses, there is emergency treatment, followed by a long recovery period, counting on nature's own restorative power. In my view the emergency action we must take is to terminate the military. Both this and the long process of behavioural modification rest on the human ability to change.

PART III
RETHINKING SECURITY

CHAPTER 6
MILITARY SECURITY IN THE NEW MILLENNIUM

The problems we face at the beginning of the twenty-first century involve interconnected issues of militarism, economics, social policy and the environment. Global consumption of resources is exceeding Earth's restorative capacity by at least 33 per cent. War and the preparation for war drastically reduce the store of these resources still further, leading to a self-perpetuating cycle in which competition for raw materials leads to further conflict. This means that global survival requires a zero tolerance policy for the destructive power of war.

However, I recognise that exposing the extremes of today's military and outlining the crisis in resources will only bring about change if we also tackle the question of security. Popular support for the military comes from fear, and that fear is based on hundreds of years of recorded history. We feel that we must have weapons to protect ourselves from the weapons of the enemy. This fear legitimises the development and stockpiling of new weapons and results in the election of public officials who will not hesitate to use violence. This in turn attracts the warrior to public office and reinforces his or her belief that military might is the best assurance of security. If the public were convinced that there were real, viable alternatives to war, such figures would lose their mandate.

Therefore it is vital that a new concept of security is devised, which puts Earth and its inhabitants first. The old paradigm of security protects wealth, financial investment and privilege through the threat and use of violence. The new concept embraces a more egalitarian vision, prioritising people, human rights, and the health of the environment. Security itself is not being abandoned; it is just being achieved through the protection and responsible stewardship of the Earth. I would call this energising new vision 'ecological security'. Such a shift in focus requires a complex, multi-faceted approach to resource protection and distribution, to conflict resolution and the policing of the natural world. In Chapter 7, I will outline some of the directions we might take towards achieving these goals. But in order to do this, we must first challenge the belief that military force is a necessary evil.

WORKING FOR CHANGE

Altering the Core Belief

Social change always follows a period when a core belief is identified and rejected. As support and awareness of this new way of thinking grows, the political climate changes and the old way of doing things is no longer acceptable. That is the lesson we learn from history. I believe, for example, that the vast social changes of the 1950s and 1960s came about when people began to challenge the idea that everyone should conform to socially imposed patterns of behaviour. This shift resulted in a new understanding of human and civil rights, with a focus on the freedom of the individual and an acceptance of racial, religious and sexual diversity.

Once a core belief is overturned, related changes spread under their own impetus. In the 1950s and 1960s we saw the growth of movements for civil rights, women's rights, black power and gay rights. Consciousness-raising in turn yields changes in legislation, social behaviour, policy, even language. More recently, we have seen the recognition of the rights of the child, the movement against child soldiers, and animal rights groups.

There will always be those who resist change – in the 1960s, the rejection of socially imposed behaviour led to fears of social chaos. But we are quick to monitor when things go 'too far' and we adjust our beliefs accordingly. So whilst we recognise the freedom of the individual, for example, this does not mean that we tolerate them violating the rights of another. Self-correction and adjustment following the rejection of a core belief is a vital part of the process.

The core belief being challenged today is that military power provides security. There exists more than enough evidence to show this belief is untrue.

There is a story about Vienna, a lovely city which was located on the path of invasion between the armies of the West and the armies of the East in the Middle Ages. The city was constantly under siege, so its warriors decided to build a high strong wall that was strengthened as time went on. At one point, however, the inhabitants began to feel crowded and wanted to expand the city beyond the confines of its wall. They had two options: knock down the wall and allow the city to grow, taking their chances with attack, or build a higher, stronger wall beyond the first. You can well imagine the heated discussions and predictions of doom that followed! In the end, the referendum was won by those who wanted the wall torn down. You can still see the remains of it today as it forms the base of the ring road around the city centre. When the wall was torn down, Vienna was no longer a challenge to invading armies, a prize to be taken, and the constant sieges ceased. This may be an oversimplification of history, but I think it underlines the need to question the proverbial wisdom at some point.

Lobbying for Change

The first step in change is the conviction that change is needed. This could be said to be the theoretical stage based on observation and reassessment. The next step is practical, when people come together to exchange ideas and information and to lobby for social transformation. What we find in reality is that these two processes occur simultaneously – discussion gives rise to groups of like-minded people who engage in further analysis.

It is clear that the multi-faceted problems outlined in this book will require a multi-faceted solution. No one person or organisation will have the wisdom needed to deal with all of the issues that must be addressed. Those working for peace, economic justice, social equity and environmental integrity must all stay connected, sharing their ideas and insight. 'Staying connected' in such a grandiose project will never mean total agreement on everything, rather a constant cycle of communication, action, feedback and re-evaluation. Honest dialogue about successes and failures is a protection against major mistakes during alternative policy development.

The good thing about such a complex range of problems is that the process can engage a wide variety of talents. Everyone should be able to find a comfortable niche where he or she can be useful and appreciated. For example, while there is a need for scientists and engineers to interpret documents, there is also a need for those who can convey the findings to the general public. The message about the dangers of nuclear weapons has been spread through art, plays, and poetry as well as through television, newspapers and magazines.[1]

Once an individual has identified the skills they have and the issue they want to address, they need to find a suitable group of like-minded people with whom they can work and from whom they can derive support. Movements that promote peace, emphasise food safety, work towards social equality and protect the environment already exist, as do organisations that continue the attack on the 'old belief' – the anti-war movements, Abolition 2000 (for nuclear disarmament), coalitions against the arms race, and campaigns against the arms trade. If no organisation or movement exists for the issues identified, the individual might need to form one. Collaborative action can involve local groups, such as school boards, professional societies, churches and service clubs as well as international organisations such as the World Wildlife Fund, Greenpeace, and Unicef. The most important thing is that these efforts must be cooperative and not competitive. The way we organise for reform is part of the solution for healing. If confrontation and competition have led to excessive greed and violence, then we require the opposite skills to rectify the imbalance.[2]

PHASING OUT THE MILITARY

So how would we actually go about bringing an end to the military? The first and most important requirement is that the military come under civilian control; then we must look at effective disarmament and the redirection of military resources, including human resources, towards more humanitarian aims; finally we must seek alternative means of solving conflict. We also need to bring the research community into this equation so that disarmament becomes a long-term reality.

Control of the Military

Many people were shocked when NATO decided to bomb Kosovo on its own authority. If NATO or some other coalition outside of the United Nations can dictate military policy then the chances of promoting a peaceful solution to any crisis are seriously damaged. There is more security for the public when international actions are based on decisions made by a civilian authority and are backed by the rule of law.

At a luncheon held in New York City on 8 July 1999, Jayantha Dhanapala, United Nations under-secretary general for disarmament, warned that the rapid globalisation of the arms industry meant that 'money, information and decontrolled commodities flow between countries and global affiliates without any significant control by governments'.[3] This freedom to trade arms means that countries or organisations can secretly build up weapons reserves, and also that highly destructive weapons can be developed, without any civilian oversight or control. An international police force under the control of the United Nations would make this clandestine build-up to war impossible. Since it would have equal responsibility to all nations and would therefore not be subject to the competition that exists between countries, it would not require constant increases in its firepower. Of course, it would have to be accountable to the UN General Assembly for its actions and decision-making and it would be important not to concentrate too much power in any one agency or department. When power is dispersed, it is less likely to be abused.

However, it is clear that the goal of change is not just civilian supervision of the military but the dismantling of the military altogether. This change will not be easy. No country is going to terminate its military forces unless it can be absolutely sure that other countries are doing the same – the fear of being vulnerable to attack would be much too strong.

Disbanding the Military

The United Nations, with the assistance of NGOs like SIPRI, has been tabulating military expenditure and arms trade transfers for many years. Enough data is now available to successfully monitor a freeze in military spending. Once a freeze is in place, 20 per cent of each country's budget could be exacted each year by the United Nations for purchase of UN currency specifically introduced for this purpose. This UN money could then be restricted to the creation of jobs that meet human needs and also conform to tough ecological and planetary health requirements:

- all production must meet a genuine planetary need;
- sustainable production, distribution, consumption, and waste disposal methods must be in place before any production begins;
- initially, jobs created must contribute directly to the health, education, social service or environmental restoration sectors of the economy.

In this way military spending is phased out, jobs for military personnel are provided, and some of the most urgent planetary needs are addressed. This financial shift could continue for five years until the military budget is zero for all member nations, although it may take up to ten years. The plan would result in the retirement of older personnel and the shifting of young recruits into a permanent United Nations Peacekeeping Force. One difficulty with this plan might be the shifting of high-tech workers into fields more traditionally viewed as low-tech. However, scientists accustomed to working with complex systems would be equally challenged by the immense environmental, health, and social questions that require attention today. Life sciences, which traditionally did not attract the physicist or mathematician, would benefit from the influx.[4]

An alternative suggestion is to redefine the military's job description. After all, they are supposed to work for us, and in our name. Proposals include using military personnel for civilian assistance in ecological crises such as floods or volcanic eruptions. They could also carry out genuine peacekeeping, with new non-violent training programmes and the development of conflict resolution skills. Imagine unarmed peacekeepers, trained in the art of diplomacy. When the option of war is not available, people are forced to think about the many possible but untried responses.[5] One of the current proposals at the United Nations is the creation of a new peacekeeping force, the Rapid Deployment Brigade, which would operate under Security Council authority. There is provision for such a brigade under Article 43 of the United Nations Charter. However, the United States is directly opposing this attempt.

The US owes as much as $1.6 billion to the UN. In late December 1999, when threatened with losing its right to vote, the US agreed to pay $926 million of its debt. It placed strings on the agreement, however, including 22 unilateral conditions, one of which was the prohibition of a UN standing army. Only $650 million of the debt has actually been turned over so far, the rest being withheld until the US's conditions are met. Pressure on the US to pay its dues without such conditions would be helpful.[6]

Some members, or former members, of the military have begun to question the relevance of their activities, such as the Retired Generals Opposed to Nuclear War, who have been so vocal in support of eliminating all nuclear weapons. Those who believed in the role of the military as the protector of women and children found that the reality was very different. The slaughter in Rwanda was a case in point:

Like many of my colleagues, I drove into [Rwanda] believing the short stocky ones had simply decided to turn on the tall thin ones because that was the way it has always been. Yet now, two years later... I think the answer is very different. What happened in Rwanda was the result of cynical manipulation by powerful political and military leaders. Faced with the choice of sharing some of their wealth and power with the [insurgent] Rwandan Patriotic Front, they chose to vilify that organisation's main support group, the Tutsis... The Tutsis were characterised as vermin. Inyenzi in kinyarwanda – cockroaches who should be stamped on without mercy... In much the same way as the Nazis exploited latent anti-semitism in Germany, so did the forces of Hutu extremism identify and whip into murderous frenzy the historical sense of grievance against the Tutsis . . . This was not about tribalism first and foremost but about preserving the concentration of wealth and power in the hands of the elite.[7]

Of course not everyone in the military takes such an enlightened view, and there is bound to be military resistance to the new concept of security. I regard NATO as one of the greatest obstacles to general disarmament in Europe and North America.

European security

NATO is trying very hard to carve out a job for itself in the post Cold War era. Although it still commands respect among those who cling to the idea that military power means security, it is growing more and more anachronistic among those whose vision has shifted. It is an alliance with no legal or juridical connection to the United Nations or global governance. In a *Guardian* article on 'truly ethical policies', Richard Norton-Taylor and Simon Tisdell stated 'Britain should press for the abolition of NATO. That organisation is a relic of the cold war: it served a purpose then, but it has outlived its usefulness'.[8] The article also advocated a reduction in British industry's dependence on the arms trade and greater involvement by the Organization for Security and Cooperation in Europe (OSCE) in solving European disputes. This would of course run contrary to the US military's desire to 'prevent the emergence of European-only security arrangements which would undermine NATO'.[9]

Indeed, as was noted in Chapter 1, NATO is expanding in Europe and the former Eastern bloc countries in a way that appears to be in competition with the OSCE. At a meeting of the Allied and Partner foreign ministers in Sintra, Portugal, on 30 May 1997, NATO formed a Euro-Atlantic Partnership Council (EAPC). According to the *NATO Review:* 'active participation in the Euro-Atlantic Partnership Council and the Partnership for Peace, will further deepen [a nation's] political and military involvement with the (NATO) Alliance'.[10] Membership of Partnership for Peace is a preparatory step to full NATO membership.

The military concept of security forms the basic tenet of NATO and there is pressure on new members to purchase the latest military hardware and software, regardless of other domestic needs or priorities:

NATO's procedures for admitting new members will ensure that the Alliance's overall goal of strengthening security for all of Europe will be achieved. The new members will be thoroughly prepared for the responsibilities and obligations of membership. In joining NATO they will join an Alliance not only committed to cooperative relations, but also open to other democracies able and willing to pursue the common cause of security and stability in Europe.[11]

New members will be obliged to accept NATO nuclear policies, reinforcing its limited interpretation of 'security and stability'. In this context, 'cooperation' refers primarily to military compatibility. According to NATO Deputy Secretary General Sergio Balanzino, the military scope of the Partnership for Peace will include:

'more complex and robust' military cooperation and common exercises, cooperation in armaments, crisis management exercises and civil emergency management, as well as commonly agreed program targets covering a 'broader range of more complex requirements'.[12]

At a meeting on 15 December 1999, NATO foreign ministers issued a promise to 'review Alliance policy options in support of confidence and security building measures, verification, non-proliferation, and arms control and disarmament, so that a comprehensive and integrated approach was secured.' Most observers consider this to be a move in the right direction, although 'diplomatically dressed [in] ambiguous language.'[13] It will only be a truly positive move if we can agree on a broader interpretation of 'security' and effectively change NATO goals.

Ending Conflict

Non-violent approaches to conflict resolution are now gaining momentum and legitimacy, with the banning of certain types of weapons, such as chemical and biological arms and nuclear weapons in outer space. However, these efforts simply narrow the scope of war rather than eliminate it. It has sometimes also been the case that these essentially positive treaties lack clear, legally binding definitions. The ambiguity in the 1972 Anti-Ballistic Missile Treaty with Russia, for example, has led to infinite legal wrangling and the development of space shields and theatre anti-ballistic missiles which may or may not violate the treaty, but will surely lead to an escalation in the worldwide arms race. The UN has failed to define 'nuclear explosion' and 'outer space' accurately. This leaves legal loopholes out of which a country can wriggle if they are intent on developing high-tech weapons.

War itself needs to be banned. There are no disputes between nations that cannot now be resolved, at least temporarily, with mandated periodical review by a court. Recent developments in international law include the launching of an International Criminal Court and well-developed plans for an International Environmental Court. With the development of effective peacekeeping forces and an investment in arbitration and mediation skills, we should be heading towards an exciting new era of real diplomacy. Indeed even after a war, negotiations are necessary before 'peace' is established. The main accomplishment of the violence is to force concessions at the negotiating table, but because a war influences the 'freedom' of the loser, post-war negotiations are notoriously unjust. Often this sets the stage for the next war – one reason, perhaps, why the Second World War followed on so swiftly from the First. With the Chemical Weapons Convention, banning chemical warfare, which came into force on 29 April 1997, the Russian Duma ratifying the Strategic Arms Reduction Treaty on 14 April 2000, and review of nuclear weapons reduction on the United Nations agenda for the same year, it seems to be the opportune moment to push this non-violent agenda.

TWO SUCCESS STORIES

Landmines

One of the most effective citizen initiatives in recent history has been the global ban on landmines. Jody Williams, who spearheaded the International Coalition to Ban Landmines, won a Nobel Peace

Prize for her efforts.

A landmine is a small, relatively cheap type of bomb. Typically it explodes when stepped on or when someone walks into a trip wire. Landmines are meant to destroy legs or arms rather than kill, because this is thought to have a greater psychological effect on the 'enemy'. In reality, landmines are usually triggered by women and children engaged in subsistence farming long after hostilities have ceased. In many parts of the world, it is children who gather firewood and water for the family, exposing them to the danger of such devices.

There are three kinds of landmine presently in use. The first is called a fragmentation mine, and these can be 'directional' or 'non-directional'. The directional fragmentation mine is usually mounted above ground, packed with steel balls or metal fragments, and placed in front of an explosive charge. It can be detonated by a trip wire or remote control, and it scatters fragments as far as 50 metres in front of its placement in a 60-degree arc. The non-directional fragmentation mine is also usually placed above ground. When someone walks into the trip wire it explodes, expelling fragments around it in a 20-metre radius. The second type of mine is called a 'blast mine'. It is usually placed just below the surface and is designed to explode when stepped on. It usually blows off the leg of an adult and kills a child. The third type is the 'bounding mine', which is also buried. Stepping on the fuse on the top of this mine, or walking into the trip wire, projects the mine out of the ground to a height of one metre or more, where it then explodes, blowing apart a child's head or seriously damaging an adult's upper body. Removing these mines is possible, though dangerous and time consuming. There is one type of blast mine, however, which cannot be touched, a rubber-covered device called a PMA-3. The cover deteriorates in the ground, leaving the device very unstable and sensitive, so it must be blown up in place.

The United Nations estimates that landmines kill or maim about 25,000 people every year. The problem is immense, with an estimated 60 to 70 million mines deployed around the world.

LANDMINES BY COUNTRY

Afghanistan	10 million	Mozambique	2 million
Angola	10–15 million	Namibia	50 thousand
Bosnia-Herzegovina	0.6–1 million	Nicaragua	116 thousand
Cambodia	4–6 million	Somalia	1 million
Eritrea	0.5–1 million	Sudan	0.5–2 million
Iraq	10 million	Other	20 million

Africa is the most heavily mined, with as many as 30 million devices in 18 countries.[14]

Removing landmines is a difficult, slow, and nerve-racking job. Greg Ainley, a 21-year-old from Edmonton, Canada, explains how it is done:

> If the grass is long, we'll definitely be taking our time. Usually it's about a metre every 10 minutes or so. We're down on our bellies, hands out in front of us feeling nice and slow with our fingers. That allows us to pick up any abnormalities that might be a mine. If a mine is found, the first step is to pull it out of the ground with a rope that stretches back to a safe place, behind an armored vehicle, for example. We pull it. That's just in case there's any booby-traps underneath. That being so, the mine will obviously detonate. If not, that's when we go over there and do our job.[15]

In 1993, 80,000 mines were removed in this painstaking way. It is challenging but satisfying work. However, during the same year 2.5 million more mines were planted. The removal of the 80,000 mines cost $100 million, and all of the resources used in the production of the devices became non-recyclable waste.

The peace movement, largely through the efforts of women, has been working to ban landmines since the early 1990s and in 1994, the International Red Cross added its voice to the protest. The campaign enlisted the help of Diana, Princess of Wales, who used her celebrity to bring the humanitarian dimension of the problem to the public, emphasising the extraordinary proportion of children killed or maimed for life. In October 1996, the Canadian government convened a meeting in Ottawa of 50 governments favourable to a complete ban, and in December 1997 some 90 countries signed a special treaty drafted in Oslo. Britain and France, major exporters of landmines, agreed to the ban but the US decided not to sign because it wanted to use the weapons in the demilitarised zone of Korea.[16] Other non-signing producers of landmines were Russia, China, India, Pakistan and Israel.

We must remember that landmines are not only the responsibility of those who lay them, but also of those who supply them. If all countries were to stop creating such weapons then the supply line would be cut off. However, as long as some countries are willing both to produce and sell landmines, this 'starvation' tactic will not hold. The treaty does not tackle the problem of the 80 million mines already planted, nor does it prohibit mines designed to blow up vehicles or disable tanks. Nevertheless, it is one small step towards phasing out the violence of war. It clearly places great value on individual lives, especially those of women and children, and it has the added benefit of protecting agricultural land that becomes useless when strewn with bombs. The ban on landmines provides a model of cooperation between non-governmental and governmental organisations, with widespread grassroots support, and gives encouragement for future initiatives.

Nuclear Weapons

The World Court Project

A second successful initiative was the World Court Project, an idea strongly promoted by Commander Robert Green, a retired British navy officer. According to Green, although there are prohibitions against weapons of mass destruction, military personnel are told that nuclear weapons have never been outlawed. Quoting from the US *Military Manual*: 'The use of atomic weapons cannot be regarded as a violation of international law in the absence of any customary law or convention restricting their use.' As a commander of ships carrying nuclear warheads, Green has always been bothered by this. Since both the US and UK military manuals require personnel to adhere to principles of international law relating to warfare, Green reasoned that a declaration of the International Court of Justice would go a long way towards eliminating these weapons and supporting military personnel who refused to use them.

Commander Green recognised the leverage that might be gained from an action of the World Health Organization, so he approached Hilda Lini, Health Minister of the Island Nation of Vanuatu and delegate to the WHO in Geneva. Minister Lini agreed to introduce a motion that the WHO seek the opinion of the World Court as to the legality of the use of nuclear weapons in war. Twenty-two nations submitted briefs in support of this motion but the US, UK, France, Russia, Australia, the Netherlands and Germany, who challenged the authority of the court to rule on such a question, opposed it. Nevertheless, the motion was passed and submitted to the World Court in 1993.

In a separate action, the UN General Assembly broadened the question to include the legality of the threat of war, as well as war itself:

> The General Assembly... decided, pursuant to Article 36, paragraph 1, of the Charter, to request the International Court of Justice to render its advisory opinion on the following suggestion: 'Is the threat or use of nuclear weapons in any circumstances permitted under international law?'[17]

Despite attempts to block this request, the resolution was passed by 78 votes to 43, with 38 abstentions. Previous attempts by non-aligned countries (those countries not aligned with either the communist or capitalist voting blocs) to table such a resolution had failed because the nuclear states had threatened trade and aid sanctions. However, because the WHO request had already been submitted to the World Court, this time the resolution was successful.

As it happened, the World Court accepted the request of the UN General Assembly, but not that of the WHO, since it reasoned that nuclear war was not a 'health' problem until after the fact. This is rather a shortsighted view, of course, since every phase in the production and deployment of nuclear weapons is hazardous to public health. However, in their ruling, the judges had no brief before them explaining the potential dangers of nuclear weapons production. Apparently the law also has no precedent for respecting the field of preventive medicine.

Retired Commander Green spoke out publicly in Europe and North America, in support of the International Court of Justice review. General Charles Homer, head of the US Space Command, also spoke out in favour of abolishing nuclear weapons. These individuals and others, such as international law expert Richard Falk, gave impetus to popular support for the initiative. In fact, the World Court motion was accompanied by intense civilian action through a coalition of international peace groups such as the International Peace Bureau, the War and Peace Foundation, International Physicians for the Prevention of Nuclear War, Women's International League for Peace and Freedom and the International Association of Lawyers for the Abolition of Nuclear Arms. This coalition activity focused on a provision of the World Court constitution that had never been used before. According to this provision, the judges are obliged to take into account 'the dictates of public conscience'. For this reason over a hundred million individuals sent in a declaration of conscience, stating:

> It is my deeply held conscientious belief that nuclear weapons are abhorrent and morally wrong. I therefore support the initiative to request an advisory opinion from the World Court on the legality of nuclear weapons.

After receiving briefs from various governments and these public statements of conscience, the International Court of Justice issued a Communique in July 1996[18] stating that:

> THE COURT unanimously, DECIDES a threat or use of force by means of nuclear weapons that is contrary to Article 2, paragraph 4, of the United Nations Charter and that fails to meet all the requirements of Article 51, is UNLAWFUL...

> Unanimously, DECIDES, a threat or use of nuclear weapons should also be compatible with the requirements of the

international law applicable in armed conflict particularly those of the principles and rules of international humanitarian law, as well as specific obligations under treaties and other undertakings which expressly deal with nuclear weapons...

By seven votes to seven, it follows from the above mentioned requirements that **the threat or use of nuclear weapons would generally be contrary to the rules of international law** applicable in armed conflict, and in particular the principles and rules of humanitarian law. However, in view of the current state of international law, and of the elements of fact at its disposal, **the Court cannot conclude definitively whether the threat or use of nuclear weapons would be lawful or unlawful in extreme circumstances of self-defense, in which the very survival of the State would be at stake.**[19] [author's emphasis]

Unanimously, DECIDED there exists an obligation to pursue in good faith and bring to a conclusion negotiations leading to nuclear disarmament in all its aspects under strict and effective international control.[20]

It is interesting that the court's support for nuclear disarmament was unanimous whilst it was split on the section dealing with 'extreme circumstances'. Observers who were actually present at the court say that this compound statement was really a political ploy so that the court did not have to deal with each part of the resolution separately.[21] For example, some judges voted 'no' because they objected to the clause: 'the threat or use of nuclear weapons would generally be contrary to the rules of international law applicable in armed conflict, and in particular the principles and rules of humanitarian law'. Other judges voted 'no' because they found the exemption for 'extreme circumstance of self-defense' abhorrent. Therefore a 'no' vote had two very different meanings. Had the two parts of this decision been voted on separately, it is likely that the first would have passed and the second been rejected.

Overall, however, the outcome was encouraging. It demonstrated that there was some support within the military for placing limits on violence, especially for banning nuclear weapons. Moreover it demonstrated that ordinary citizens could successfully engage international organisations like the World Court. It was heartening to see that both governments and the public respected this legal intervention to limit weapons of war. This second success story, like the first, involved collaboration between individuals, governments and international organisations.

The Canberra Commission

The Canberra Commission was an independent international commission of 17 experts on war and peace, established by the Australian government in November 1995. Its goal was to define practical steps towards a nuclear-free world. The commission held four meetings over the course of a year and issued its first report in August 1996. It has developed practical plans for systematically ridding the planet of nuclear weapons.

Canadian Senator Douglas Roche, Canada's former Ambassador to the United Nations on Disarmament, has also been putting forward the so-called Middle Power Initiative, whereby medium-sized nations such as Canada or Sweden could be called upon to mediate disputes involving the major nuclear powers or NATO.

These two efforts flow from and strengthen the decisions of the World Court. They demonstrate the simultaneous breaking down of the old world order and the creation of a new energy that will provide the impetus towards greater global security.

The UN response to the Advisory Opinion of the World Court

The historic decision of the World Court provided members of the UN with the necessary legal backing to push once again for nuclear disarmament. Many of the developing nations that had been restricted by the threat of aid sanctions could now begin to make real progress. In particular, the judgment called to attention the Nuclear Non-Proliferation Treaty 1968, which bound nuclear nations to disarm.

Brazil introduced a resolution to the UN on a Nuclear Weapon Free Southern Hemisphere and adjacent areas to strengthen the international spread of Nuclear Weapon Free Zones (NFZs). NFZs had been declared by municipalities, countries and regions for many years in an attempt to protect them from nuclear aggression. Some of the best-known NFZs are South America and New Zealand. The Brazil resolution was designed to 'imprint upon the public conscience the image of a globe already free from the scourge of nuclear weapons over more than half its surface'.[22] This resolution was amended by Pakistan to include South Asia. It was passed by 111 votes to 4, with 36 abstentions. The subsequent nuclear weapon testing in India and Pakistan was doubly regrettable after this vote.

Malaysia introduced into the First Committee of the 51st UN General Assembly a draft resolution welcoming the Advisory Opinion of the Court and calling on all states to fulfil their obligations by 'commencing multilateral negotiations in 1997', leading to a nuclear weapons convention. On 26 November 1996, this Malaysian resolution was adopted with a vote of 94 in favour, 22 opposed, and 29 abstentions. China voted 'yes', while the US, UK, France and Russia voted 'no'. Other 'no' votes included Canada and the other European countries, with the exception of Ireland and Sweden. Australia and Japan abstained. The Non-Aligned Movement then proposed a step-by-step plan for reductions leading to total disarmament in a time-bound framework. This was also adopted by a vote of 87 to 38, with the objection of the Western nuclear and European states.

The United Nations and the Non-Governmental Committee on Disarmament have now undertaken negotiations leading to a Nuclear Weapons Convention which would prohibit the development, production, testing, deployment, stockpiling, transfer, threat or use of nuclear weapons. The convention calls for the nuclear weapon states (US, UK, France, Russia and China) to make 'systematic and progressive efforts to reduce nuclear weapons globally, with the ultimate goal of eliminating these weapons'. While negotiations continue, it looks promising that nuclear weapons will be banned by the year 2020.[23]

While this may seem slow to those who have not slept well since 1945, it is a major blow to militarism. The Carnegie Endowment for International Peace reported that in 1995, the five nuclear powers possessed 36,816 nuclear weapons. These continue to pose a serious threat to security even though the Cold War is over.

TAKING ON THE RESEARCH COMMUNITY

Although the destruction of current weaponry is a primary aim, long-term demilitarisation requires a change in the way scientific research is funded. Money for military research comes from governments or from the business community. Grants are awarded to universities or students investigating areas of interest to the military, and multinational corporations are 'excused' from taxes if they devote money to high-tech research. As long as this is the case, it will be difficult to shift funds away from military towards domestic needs. If we are to pursue a new form of global security, then all national research priorities must be put under civilian control, and that research must be subject to public scrutiny. In this way academic freedom can be protected and simultaneously freed from the exploitation that would use it for destructive or violent purposes.

Some possible mechanisms for controlling and redirecting research come to mind:

- an International Research Council (IRC), which would review all research proposals requiring funding in excess of $50,000;

- an international academic review panel in each major discipline, as part of the IRC, for handling all research grants exceeding $50,000, with a yearly change in panel membership;

- a requirement that each large research grant proposal provide built-in mechanisms to thwart any military use of the findings, including open publication of all results;

- a requirement that all research proposals of value higher than $50,000 be interdisciplinary and involve at least one partner from both the ethical and social sciences, as well as research team members from more than one country;

- international publication of all funded research, similar to the current annual publication of research projects in cancer;

- collaborative research between universities in the northern and southern hemispheres.

It is likely that NGOs capable of monitoring compliance and enforcing these global research regulations would spring up, as they have done for arms control and human rights. Again, what is most important here is transparency of purpose, cooperation and an interdisciplinary approach.

THE IMPORTANCE OF GRASSROOTS ACTIVISM

Although ordinary people do not have direct access to the International Court of Justice or the UN General Assembly, they are effectively influencing these bodies in a very profound way. Reflecting on the World Court decision, Fredrik Heffermehl of the International Peace Bureau stated: 'This case is an encouraging example of the ability of people's organizations to make use of international institutions like the World Court, which are meant to serve the world's people and not only their governments.'[24]

The global popular consensus against nuclear weapons is remarkable. There is international support and a great deal of grassroots activism working vigorously towards total nuclear disarmament. It is astounding that governments should be so out of touch with popular opinion in countries claiming to be world leaders in democracy.

The power of this unofficial international coalition was again demonstrated when 10,000 grassroots activists from more than 80 countries travelled, without international funding assistance, to the Hague Peace Conference in 1999. If each of those present represented 100 other activists who were unable to travel to the Netherlands, this would conservatively place the size of the global coalition at a million people. Many of the same activists were present on the streets of Seattle in November 1999 to campaign against the policies of the World Trade Organization, and in Washington DC in April 2000 to protest at the International Monetary Fund and World Bank's failure to meet the needs of the world's poor. In Seattle the street protests were matched by the dissension of representatives from developing countries participating in the conference. Since the Washington meeting comprised only the seven leading developed nations, this synergy was not observable.

It is certainly reasonable to believe that the energy for change is increasing and that it is being promoted by a significant coalition of non-governmental organisations in developed countries and governmental organisations in developing countries. It will be important to direct this energy in a constructive way. There is evidence, as has been pointed out in Part 2 of this book, that warrior nations have already abandoned most types of nuclear arms and have moved towards the use of various electromagnetic beam and pulse weapons. Many of the original peace movements have disbanded and others have turned their attention to the dismantling of nuclear weapons. I imagine that some military strategists may be quite pleased with this situation.

At present organisations to prevent war in Earth's upper atmosphere are not as well organised as the anti-nuclear movements. However, the issue of electromagnetic weapons, plasma shields and laser beam weapons may be amenable to similar initiatives. The Cassini rocket, which sent plutonium into space, has so far attracted the most public attention. There was a major rally in Washington, DC on 14–17 April 2000, called 'Keep Space for Peace'.[25]

The time is also ripe for a feminist critique of human and environmental security needs and the conditions that must prevail in order to support those needs. The feminist view has been lacking in international affairs for a long time, yet this perspective is central to the restoration of peace and diplomacy. There is an undisguised assumption by those trained in military arts that since they protect women, women owe them something in return. Feminist researcher Betty Reardon has identified five categories of the abuse of power which have given rise to women's resistance to militarism: military abuse of civil society; abuses within the military; abuses of military power; dereliction of public responsibility; and military violence against women. 'Militarization connotes the over-privileging of the military and the application of military authority to political issues and problems,' she has written, and she describes a paradigm shift from military-based to human-based security.[26] I would go still further and call for ecological security.

A feminist critique would also explore the issue of culturally sanctioned violence. The hard sell marketing of military and space to young people is apparent in all forms of media. Violence intrudes into our homes through television and video games in which the player must act quickly and without thought in order to 'zap the enemy' on command. This, of course, is far removed from the reality of war, but it does reinforce the notion that armed conflict is acceptable and does little to stimulate a child's intellectual development.

Ordinary people offer two important things the military requires for its existence: legitimacy and resources. Resources include natural resources, human intelligence, money and young recruits. If we begin to see war as an illegitimate means of solving disputes and withdraw our resources, it will send a strong message to those who define policy. Our social institutions and political organisations can also form part of this movement for change. Churches can promote the issue of non-violence and military chaplains can raise consciousness among military personnel. Schools must teach peace rather than war as the most dynamic force of history, and children and adults alike can learn how to resolve problems without resorting to violence. Each one of us is important in this process, and each one can find a way to pitch in for peace.

A DECADE FOR NON-VIOLENCE: 2000 TO 2010

In September 1997 the International Fellowship of Reconciliation together with 20 Nobel Peace Prize laureates appealed to all members of the UN General Assembly to declare the first decade of the new century a 'Decade for a Culture of Non-violence'. The decade will focus on eliminating the physical violence, psychological violence, socio-economic violence, and environmental violence that victimise children all over the world.[27] Children are brutalised by the displays of aggression that occur in schools, on the streets, in families, and in the community. Think of what the children exposed to the conflicts in Rwanda, Iraq, and the Baltic states must have witnessed. They are deeply wounded by these experiences and often grow up to repeat them.

The time appears to be ripe for such a concept. War and violence have been the answer to perceived injustice and unregulated desire for power, property, and wealth for too long. We need to think beyond the old paradigms and allow ourselves to imagine the alternative. The Decade for a Culture of Non-violence opens with a UN review of the Nuclear Non-Proliferation Treaty – the opportunity to extend non-proliferation to all war and weapons of mass destruction should not be lost.

CHAPTER 7
ECOLOGICAL SECURITY

Boutros Boutros-Ghali, speaking at the Summit for Social Development in Copenhagen in 1995, argued that the concept of security on which the United Nations was built, namely preventing an aggressive attack by one nation on another, was now almost non-existent.[1] He described a 'new crisis in human security', characterised by increasing internal conflicts, mass migration, an increase in the number of urban slums, rising social tensions, psychological distress and disease, international drug trafficking, organised crime, over-consumption and pollution.

It is my belief that the best way we can ensure security for future generations and for Earth itself is through a sound combination of economic, health, environmental and social programmes. Links are now being made between air, water and food protection and the existence of a sound economy and a healthy community. No part of this balance can be sacrificed without losing all three. Unicef has noted a growing consensus that to meet these new security needs we must distribute the benefits of society equitably: regenerating the environment rather than destroying it; empowering people rather than marginalising them. Policies in the future must be pro-poor, pro-nature, pro-democracy, pro-women and pro-children. What a marvellous checklist for the assessment of new projects!

I have singled out the military as the key to quick 'surgical' action; trying to resolve the social and environmental problems that are often directly related to their activities may take much longer. As people begin to realise that some of the things happening to Earth are side effects of deliberate experimentation and a misguided set of priorities, they may respond with anger. It is important that this anger is transformed into positive energy and focuses on how to improve our situation in a practical way. As with phasing out the military, the introduction of ecological security and sustainable development will require a cooperative, interdisciplinary approach, the sharing of expertise and ideas, and action on both global and local scales. Again, we must proceed with caution, employing self-correcting policies and taking note of our successes and failures.

The pooling of experience is a hugely important resource when it comes to diagnosing a problem and healing it. An awareness of the illnesses of soldiers exposed to Agent Orange in Vietnam, for example, would have helped the residents of the Niagara Falls, Love Canal area to understand the symptoms they were experiencing (see p157). Most large environmental organisations are now fully incorporated into the structure of the United Nations, and other agencies can benefit from their insight. Eventually the many organisations campaigning against the military will need to join this 'pool of knowledge' both to enrich the understanding of military pollution and manipulation of the environment, and also to enable other agencies to respond appropriately to their peace needs.

Harm to an individual or to the environment has almost invariably been silenced to protect 'national interest' – usually national security or tourism. Joining worldwide organisations and moving these local problems to the international level may be helpful in gaining unbiased attention. With the formation of global villages, national loyalties and rivalries will need serious re-examination. We become citizens of Planet Earth, still loving our unique part of the globe, our own language, culture, literature and art, but without the competition and rivalry that have dominated the past. It was the Japanese survivors of Hiroshima and Nagasaki who first taught me the value of reaching beyond the struggles of one's own nation. They invited me to the commemoration ceremonies that took place in the summer of 1978. I was able to see for myself that the suffering of the Japanese people was the same as that of the American occupation forces, the veterans of atomic testing, and the Marshall Islanders who lived near the nuclear test site at Bikini Atoll. I introduced the Japanese survivors to some American veterans, and it was touching to see them invite these 'enemies' to come to Japan as their guests. The Japanese honoured their guests and new bonds of friendship, based on their common suffering and stronger than the hatred of war, were formed. We do not need to be mutually destroyed in order to discover our common humanity. The Japanese victims of Hiroshima and Nagasaki have never expressed vengeance. They have become radical pacifists, among the first to embrace global citizenship.

I have therefore chosen in this last chapter to outline the global structures now in place to deal with the complex problem of sustainable development and the groundwork that has already been achieved. I will look at how this global activity can be linked up to more localised concerns and action groups. It is worth reiterating the fact that questions of the economy, the environment and public health are intimately linked in innumerable ways. Measures to protect the environment should result in improved human health, better health will have a beneficial effect on the economy, and sound industrial practice will dramatically improve the security of the land on which we live. Although the overview of military abuse has been depressing, the many initiatives now in place testify to the enormous human potential for innovation and healing.

PROTECTING THE ENVIRONMENT AND SUSTAINABLE DEVELOPMENT

An Earth Charter

The Human Rights Covenants, promulgated after World War II, outlined the behaviour expected of one human towards another. Although some individuals or groups may acquire power over others – through democratic election, the seizure of power in a coup, and so on – there are limits beyond which this 'power' should not extend. A legally arrested person, for example, should not be tortured or kept in inhumane conditions in any nation. To do so would violate his or her human rights. The international court at the Hague can prosecute leaders who fail to observe these covenants.

Our human rights Covenants deal primarily with the relationship between governments and their subjects. They do not articulate the rights of animals, living creatures, or of the Earth itself, yet having such an articulation would set boundaries on what those with 'power' could legitimately undertake. Secondly, there are no guidelines as to regulating the punishment meted out to nations that fail to meet our humanitarian or economic requirements. Boycotts, sanctions, intrusive investigations, bombings, denial of food and medicines, and denial of air space all threaten the health and quality of life of people living within the country being punished. According to the United Nations, there has been an increase of 90,000 deaths a year (250 each day) in Iraq due to the sanctions.[2] During the 1990s the UN Security Council placed 12 countries under economic, trade and arms embargoes.[3] A New York-based International Peace Academy released a study, 'The Sanctions Decade', which noted that in most cases the sanctions failed to achieve their objectives.[4]

In preparation for the 1992 Earth Summit in Rio, many efforts were made to produce an 'Earth Charter' which would articulate expected human behaviour towards Earth and all living things. While a global consensus to eliminate weapons of mass destruction has been comparatively easy to obtain, agreement on acceptable behaviour towards the environment has been much more difficult. Cultural and religious practices range from Buddhist protection of the lowliest insect to the Christian doctrine that all of Earth's resources are for the service of humans, the 'highest' creation. Therefore by the time of the Rio summit, almost 100 different Earth Charters had been proposed, some quite lengthy. Yet it was also clearly recognised that an Earth Charter was not only desirable but vitally important to the future of the planet. The economic development model being followed throughout the world was exploitative of the Earth and could not be sustained.

Subsequent to the Rio summit, a collection of international NGOs began to sift through all the proposed Earth Charters in order to devise a document that would harmonise the ideas, was relatively short, and could form the basis of a global consensus. Indigenous peoples' organisations, women's networks, youth and faith groups, universities, schools, consumer groups, trade unions, peace groups, business organisations, local governments, scientific organisations, environmental groups, research centres, United Nations agencies, global governance organisations and media networks were all consulted. Many small organisations and individuals devoted themselves to this task without any remuneration (as is typical of most worthwhile international activities). Finally, in time for the Rio +5 Conference, there was a draft document ready for consideration.

THE EARTH CHARTER
Benchmark Draft II, Abbreviated Version October 1999

TOGETHER IN HOPE WE PLEDGE TO:

1. Respect Earth and all life.
2. Care for the community of life in all of its diversity.
3. Strive to build free, just, participatory, sustainable and peaceful societies.
4. Secure Earth's abundance and beauty for present and future generations.

IN PURSUIT OF THESE GOALS, WE WILL:

5. Protect and restore the integrity of Earth's ecological systems, with special concern for biological diversity and the natural processes that sustain and renew life.
6. Prevent harm to the environment as the best method of ecological protection and, when knowledge is limited, take the path of caution.
7. Treat all living beings with compassion, and protect them from cruelty and wanton destruction.
8. Adopt patterns of consumption, production, and reproduction that respect and safeguard Earth' s regenerative capacities, human rights, and community well being.
9. Ensure that economic activities support and promote human development in an equitable and sustainable manner.
10. Eradicate poverty, as an ethical, social, economic and ecological imperative.
11. Honor and defend the right of all persons, without discrimination, to an environment supportive of their dignity, bodily health, and spiritual well being.
12. Advance worldwide the cooperative study of ecological systems, the dissemination and application of knowledge, and the development, adoption, and transfer of clean technologies.
13. Establish access to information, inclusive democratic participation in decision making, and transparency, truthfulness, and accountability in governance.
14. Affirm and promote gender equality as a prerequisite to sustainable development.
15. Make the knowledge, values, and skills needed to build just and sustainable communities an integral part of formal education and life long learning for all.
16. Create a culture of peace and cooperation.

The entire EARTH CHARTER can be found on the internet: *http://www.earthcharter.org/draft/*. Education materials related to the EARTH CHARTER are available from Global Education Associates, 475 Riverside Drive, Suite 1848, New York NY 10115 USA.

Some of the poetry and balanced structure of potential Earth Charters had been lost in the amalgamation, but the proposed document, further modified at the conference, was unanimously accepted by the 700 delegates representing governments, NGOs and business organisations from over 80 countries. It is being circulated to the member states of the United Nations, and further refined between June 1997 and December 2002 in an official review process. It has been well received and signature is anticipated in December 2002. Its ratification will be accompanied by recommendations to member governments as to its enforcement within their national boundaries. If successful, the Earth Charter will become the foundation on which international law on the environment is built.

The Earth Charter has specific implications for the future of the military. It promises to 'strive to build free, just, participatory, sustainable and peaceful societies' and to 'create a culture of peace and cooperation'. It leaves little room for industry to produce harmful chemicals such as Agent Orange since all countries are bound to 'protect and restore the integrity of Earth's ecological systems, with special concern for biological diversity and the natural processes that sustain and renew life'. It is also difficult to see how one nation can impose crippling sanctions on another if it has pledged to 'honor and defend the right of all persons, without discrimination, to an environment supportive of their dignity, bodily health, and spiritual well being'.

World Governing of the Environment

New organisations are expected to form to monitor compliance with the Earth Charter, and something akin to the Human Rights Tribunal will be established to supervise the legal enforcement of its conditions. An environmental court could adjudicate on problems such as the use of depleted uranium in Iraq and Yugoslavia, or the health problems caused to those living near weapons testing sites. With closer monitoring, it might be possible to identify extreme weather events that are thought to be direct or indirect effects of human activities. Trade in illegal drugs, toxic and radioactive waste disposal, unhealthy or genetically engineered food, and the unlimited exploitation of natural resources for unneeded commodities, would all be potential cases for this new institution.[5]

Power differences between countries, classes, genders and races promote conflict rather than peace. Equal standing before a court provides a level playing field, the opportunity for an individual to take on a government, or an NGO a transnational corporation. Although the International Court of Justice did establish an ad hoc chamber to deal with environmental matters, it could not fulfil this egalitarian role as its statute, Article 34, prohibits it from hearing contentious cases involving international organisations, NGOs, multinational or transnational corporations and individuals. Under the present system, only governments can bring suit or stand accused. Even then, they can avoid an accusation by challenging the competence of the court to adjudicate on the particular issue. Multinational corporations only accept self-regulation, which almost never takes place, and often excuse themselves from good environmental practice for reasons of 'economic growth' and 'market competition'. With the establishment of the World Trade Organization and the consolidation of global capitalism, it is more important than ever to provide a strong voice for human and environmental rights. At the moment, information and publicity are the only tools available to individuals and NGOs to highlight environmental abuse.[6] As noted earlier, even access to the media can be difficult.[7]

The idea of creating an international court of the environment first surfaced in Rome in February 1988 as a private initiative put forward by experts from 30 countries, who imagined it would have only moral rather than legislative authority. Judge Amedeo Postiglione, a judge at the Supreme Court of Italy in Rome and formerly an environmental lawyer, spearheaded this effort from the beginning.

In June 1994 the Fourth International 'Toward the World Governing of the Environment' Conference was held in Venice. This conference observed that the environmental crisis had deepened since the 1992 Rio summit. Agenda 21, the 'action plan' accepted at Rio, had provided for new permanent institutions to protect the environment, but in Venice it was noted that no progress had been made on this issue. The Venice Declaration arising from the 1994 conference decided:

'• to ask the Italian Government and the other Governments to officially support the project for an International Court of the Environment and an International Environmental Agency;
• to ask the Italian Government to set up a Permanent Committee, representing all Continents, whose headquarters would be in Venice;
• to ask that the Permanent Committee study existing means and identify immediate steps to be taken for ensuring international control and adjudication of global environmental problems, namely at governmental level, and to draft a Protocol for the establishment of an International Court of the Environment.'[8]

The International Environmental Agency would build on the work of the Earth Charter in defining international standards to which nations should comply.

Large delegations of citizens from many UN countries are now forming organisations in support of these proposals. Women have been especially vocal in calling for a just and healthy planet and the right to safe food, clean air, and uncontaminated water. Women's Planet[9] in Italy and Women's Environment and Development Organisation[10] in New York are leading organisations that welcome affiliates. The governments of Argentina and Austria have also lent their support to the idea of an international court. Parallel work is being done within the United Nations, through its International Law Commission, to define environmental crimes within the framework of the Draft Code of Crimes against the Peace and Security of Humankind. On 26 May 1993, the UN Security Council decided to set up an 'international crimes' tribunal at the Hague. The extent to which its jurisdiction might, on an incidental basis, include environmental crimes is unclear. Some are promoting this as an alternative to an International Court of the Environment. However, the skills needed for adjudicating on criminal cases and environmental cases differ, as the latter require the collaboration of technical and scientific experts to clarify some of the issues.

For persons wishing to help in the formation of these important institutions, there are national organising committees in Argentina, Belgium, Canada, Colombia, Costa Rica, France, Germany, Greece, Japan, Luxembourg, Mexico, Portugal, Spain, the United Kingdom and the United States. The number is growing.[11] Women in the United States may face more difficulties than most and might fund UNIFEM, the United Nations Organization for Women's Issues, a better forum for support. In December 1999, when the US Congress decided to pay a part of its United Nations dues, it stipulated that the money could not go towards reducing debts from the 1992 Rio summit or the 1995 Women's Conference in Beijing. None of the US money can be used for the Framework Convention on Global Climate Change, the International Seabed Authority, the Desertification Convention or the International Criminal Court. This is extraordinarily regressive behaviour for a global leader.[12] Under President Bush, the US had expressed this same dislike for environmental policy reform at the time of the Rio Earth Summit. The US has not complied with the Kyoto protocol, which arose from Rio +5 and called for reduced carbon dioxide emissions, even though its emissions are higher than any other country in the world. An International Court of the Environment would provide a means of challenging these rather blatant US policies.

Earth Council

In September 1992 the Earth Council, an NGO derived from the Earth Council Foundation, was formed. The government of Costa Rica, known for environmental sensitivity and also a commitment to peace (it has no army), invited the council to establish its secretariat in San Jose and has generously supported this new organisation.[13] The Earth Council is governed by 21 members drawn from the international politics, business, science and non-governmental communities. The original board was chosen by an organising committee, and methods of democratic replacement and transparency of decision-making are still being developed. An Honorary Advisory Board of 15 eminent world leaders has also been chosen. The organisation did not begin to function globally until 1996, so not all of its policies are yet in place. In the years since its formation, however, the Earth Council has formed many partnerships and has established six core programmes of action in support of Agenda 21. The stated mission and objectives of the Earth Council are:

To support and empower people in building a more secure, equitable and sustainable future.

Earth Council has three fundamental objectives for its work:

- to promote awareness and support for the needed transition to sustainable and equitable patterns of development;
- to encourage public participation in decision making;
- to build bridges of understanding and co-operation between civil society organisations and governments world wide.[14]

Membership of the Earth Council is open to corporations as well as NGOs. The goals of the organisation are certainly admirable, but the inclusion of industry is not unproblematic. Non-governmental organisations, which have been the partners of the United Nations from the beginning and are mostly run by volunteers, are afraid that their influence will be eroded by large, well-financed corporations. The 'heavy' voice of industry could prove the downfall of the UN as a people's organisation. (See Appendix A for current Earth Council members.)

Of course, it is also obvious that changes in economic behaviour must involve big business as well as medium and small enterprises. My own observation at the Rio +5 Conference was that Business for Sustainable Development, the International Chamber of Commerce and the World Bank delegates did not integrate well with the government and NGO delegates. They gave the impression of only being there to advise the assembly rather than to engage in serious dialogue. Perhaps in the future, experienced NGOs could tutor businesses on co-operative management practice, as the commercial sector has always been based on a competitive model.

The six programme initiatives of the Earth Council cover most of the important areas of economic reform and change in development strategy identified by both the Rio summit and Rio +5 Conference.

The first programme is the continuing development of the Earth Charter. The second will address economic reform, with improved assessments of the success or failure of economic policy, and the removal of many of the inefficient economic subsidies which have distorted development in the past.[15] This programme will involve banks, businesses, governments and NGOs who will work on restructuring economic policy on local, national and global levels.

The third programme is devoted to strengthening the involvement of civil society groups in the decision-making process. Ordinary citizens must be encouraged to become involved in growth and development projects, because local change is the key to harmonised global change. The fourth programme links the sustainable development initiatives to the abolition of war, essential if our planet is to survive. This programme will develop mechanisms for mediation and conflict resolution and includes plans for an ombudsman to redress injustices in the areas of public health, environment, development and human rights.[16] I believe it is important that this programme should build in a method of obtaining swift and accurate feedback on the impact of policy changes so that it can correct any wrong decisions.

The fifth programme of the Earth Council is a partnership linkage through the Internet, primarily through a dedicated Earth Network for Sustainable Development. In this way, an enormous number of people throughout the world could quickly gain access to new data and intervene with corrective information viewed from their own perspective. I think that this particular Earth Council strategy would provide an excellent place, for example, for veterans who suspect that they have been harmed by weapons used in war to bring their concerns. The Earth Council, not fettered by a national agenda, could provide neutral evaluation of the issues. Obviously, the data organisation, accessibility and non-censoring of such an undertaking will be crucial to its success. Also, it will be essential to devise ways of identifying the sources of information so that fact could be differentiated from speculation and so that the network would not be vulnerable to misleading or incorrect information posted by those who do not share its aims.

The final programme of the Earth Council involves the participation of 'special constituencies' in each of the other five programmes. These constituencies include indigenous people, women, youth and faith groups who rarely have an equal voice in national and international affairs. Perhaps this is because the dominant forms of communication and consultation were devised by white men, for white men. This is the problem encountered by women who try to enter the boardrooms of large companies; they are welcome if they are able to act like 'one of the boys'. Indigenous people accustomed to consultations organised in community circles, with the floor held by the person holding the 'talking stick' and agreement being reached through consensus, are unlikely to feel at home with *Robert's Rules of Order.*[17] (One encouraging development noted at the Rio +5 Conference was the new policy of the World Bank of involving local NGOs from the requesting country in the planning of major projects to be funded by the bank. The bank has also become involved in funding these organisations, a policy called 'capacity building', so that they can fully participate in the bank's negotiations for approving loans.)

The problems foreseen for the Earth Council should not be a deterrent to its formation and growth. It is a new vehicle for a new age. If it is adequately supported and encouraged, members can build on these foundations until it is a powerful tool for change.

How to Link Locally with these Global Initiatives

Ordinary citizens working on very specific local projects form the backbone of social transformation. For example, in response to health concerns and in an effort to reduce costs, farmers in Ontario have reduced their pesticide usage by 25 per cent in the last ten years. Their goal is to decrease usage still further, by another 50 per cent in the year 2002. Farmers have organised the recycling of pesticide containers and the safe disposal of unwanted or deregistered products. The Ontario Farm Environmental Coalition has developed an analysis of 17 key environmental issues facing farmers and is calling for a complete Environmental Farm Plan.

Local initiatives can gain information and support from larger national or global programmes. The support network for such projects is slowly being built up through the International Council for Local Environmental Initiatives (ICLEI), also known as the

International Environment Agency for Local Governments, which is dedicated to the prevention and solution of environmental problems through local action. In 1998 there were 300 cities affiliated throughout the world, and the number is growing rapidly.

ICLEI was launched in 1990 under the sponsorship of the United Nations Environment Program, the International Union of Local Authorities, and the Center for Innovative Diplomacy, and it has official status with the United Nations through which it can voice the interests of local government. The ICLEI provides an umbrella organisation in which individuals can work on the particular issues in their own area, yet benefit from the cumulative knowledge of other projects going on around the globe. Ultimately, it provides a way for local organisations to have their own impact on the development of innovative policies.[18]

There are currently two programmes under ICLEI: the Cities for Climate Protection Campaign and Local Agenda 21 Initiatives. A Local Agenda 21 Initiative is based on the extensive list of 'things-to-do' that arose from the first Rio conference. Those organisations involved select what seems most relevant to the needs of their community, and action plans are developed and then implemented. There are also National Agenda 21 Initiatives that provide training in areas such as environmental auditing, issue assessment, strategic energy planning, and environmental budgeting. The Climate Protection Campaign helps cities to achieve their local carbon dioxide reduction goals. It operates a variety of technical assistance projects that focus on innovative approaches to financing and implementing energy-efficiency measures in municipal and commercial buildings; reducing greenhouse gas emissions through effective waste management programmes and land-use planning; and developing strategies to reduce emissions in the transportation sector.

One example of a fruitful local initiative, linked to the global picture, is the production of an ecological footprint for the city of Toronto, using the same methodology as was used for the national footprints described in Chapter 5.[19] Through this analysis it was discovered that the largest amount of resource wastage occurred in the transport sector. The footprint also led to projects as diverse as rooftop gardens, urban agriculture, and more purchasing of local farm produce. An enterprising German scientist, Hans-Peter Durr, has developed a computer program that will calculate ecological footprints for individuals, based on their occupation, lifestyle choices and savings. He noted that people living in a very ecologically sound way may actually find they have large ecological footprints – this is because the money held in their bank accounts is being invested in projects which waste or destroy natural resources.[20]

Water and Soil Quality Assessment at the Former Clark Air Force Base in the Philippines 1997

Arsenic was detected above World Health Organization standards in six operational wells. The pesticide Dieldrin was detected above WHO standards in four operational wells, back-up wells and decommissioned wells.

One decommissioned well contained dissolved solids, sulfate and coliform bacteria above WHO standards. Ten volatile organic compounds were detected above WHO standards in one decommissioned well.

In the evacuation Camp Well, nitrate, mercury and coliform bacteria above WHO standards were found.

Pesticide levels exceeding by factors of ten to thirty the industrial soil criteria were documented; total petroleum products at very high levels (above US government clean-up levels), high levels of dioxins and poly-aromatic hydrocarbons were all found on the site. High concentrations of lead were found in the vicinity of the battery shop, more than twice the industrial criteria.

Jet fuel, benzene, toluene, ethyl benzene, and xylene were found in the aviation area.

There were two toxic waste dumps on the site, and a bombing range containing unexploded ordnance.

Prepared by Weston Industrial, West Chester, PA, USA, August 1997.

Citizens who become involved in any of these local and regional efforts will probably, at some stage, become aware of the military excesses that threaten the ecosystem. In fact, many may have to deal directly with local pollution caused by the military. As mentioned in the previous chapter, cleaning up former military bases is now a major problem. The Clark Air Force and Subic Navy bases in the Philippines are a case in point.[21] The Filipino people trying to live on the former Clark Base are now experiencing unusual sickness.[22] Although the US General Accounting Office said that the contamination of these bases would have been of Superfund level if they had occurred in America,[23] the US government is claiming no responsibility. It says its contract with the Philippines, dated 1898, did not include clean-up. Organisations such as the Stockholm Environment Institute and SIPRI compile information that can help local areas see such problems within a global framework.

It is not necessary for each small community to reinvent the wheel. The networking possible through organisations such as the ICLEI means that each community has access to an immense wealth of information and can gain insight from the experiences of other groups. As we work within our local milieu, we are contributing to the health of the whole planet. Remember the strategy: Think Globally and Act Locally.

PROTECTING HUMAN HEALTH

Preventing the effects of environmental degradation on human health – in particular child health – is a fundamental component of sound environmental policy. Human illness related to environmental pollution is one of the most wasteful and tragic dilemmas of modern living. In the popular mind, cancer is probably the most feared side effect of pollution, but other chronic diseases and unnecessary congenital malformations are equally debilitating.

Cancer rates seem to have increased consistently over the last century. This could, in part, be due to the longer life span of people in the developed world and also to the discovery of antibiotics, which reduced the rate of deaths from infections. Basically we live longer and are less subject to other diseases, so there is more 'scope' for cancer development. However, there is also a very strong link to the dramatic increase of man-made chemicals and radionuclides in the environment. Most toxic chemicals and all radionuclides are mutagenic and carcinogenic – that is, they are capable of changing DNA, the genetic (or cytogenetic) material which instructs a cell to produce some enzyme or hormone or which controls the growing and resting periods of the cell. A cancer is a runaway growth, a cell unable to rest. However, it is not always easy to trace a specific illness to its environmental cause. There is always some time period between exposure and the development of a clinically diagnosable disease. Also, the average person is exposed to so many chemicals, what is called the 'toxic soup', that it is almost impossible to sort out the effect of each separately.

Bearing that in mind, it is clear that cancer rates are not the best indicators for measuring the effects of pollution, yet they are often used in studies of public health. Better measurements would be episodes of human respiratory distress per day or week, the rate of human or farm animal miscarriage, bird or fish kills, or the rate at which bacteria mutate and become drug resistant. These indicators give clear early warning of environmental strain. We also do not have good historical resources as to the effects of environmental pollution. Our current system for keeping records of public health is crude and outdated, as it still concentrates on infectious diseases and food poisoning. In the developing world public health policy focuses on problems such as malaria, typhoid, and tuberculosis, so the effects of illness caused by toxic pesticides and other imported chemicals go largely unrecognised. Reproductive abnormalities, neuro-toxicological effects, chronic illness, cancers and genetic disorders are now more relevant indicators of the general health of the population. Health monitoring of these factors is very different, so major changes in the way data is collected and interpreted are urgently needed.[24]

We can also glean relevant information about environmental pollution by considering its effects on animals. As noted above, if the fish population of a lake dramatically decreases, it is a strong indication that there may be serious problems in the local area. We are all part of the biosphere and the health of one part affects the whole. On a very simple level, if humans eat contaminated fish from the lake, their health may suffer too. Good animal health is as vital to the well-being of the planet as clean air and water.

In 1993, members of the Great Lakes Research Consortium[25] conducted a telephone interview with coordinators at suspected pollution 'hot spots' around the Great Lakes. They compiled a list of the health concerns that were mentioned by ordinary people living in the problem areas. Thirty-two per cent mentioned air quality; 29 per cent drinking water; 21 per cent cancer rates; 11 per cent contamination of the fish; and 7 per cent reproductive health. In response to this public concern the Great Lakes Human Health Effects Research Program prepared two guide books for the public: *Investigating Human Exposure to Contaminants in the Environment: A Community Handbook* and *A Handbook for Exposure Calculations*. They are also in the process of receiving feedback on a 400-page draft of a *Handbook for Health Professionals: Health and Environment.*[26] These are excellent basic texts that can be used to empower any community and support local initiatives.

The Great Lakes example highlights the need for public access to the uncensored results of any studies carried out into environmental pollution. Women in the Great Lakes Basin have been quite vocal about their distrust of the way government agencies and industry word their statements about the environment. 'Positive messaging' together with military secrecy can be misleading about the severity of environmental problems. Admittedly, there is a fine line between scaring the public with information and encouraging them to take action. Communicating scientific information, especially when all of the questions are not yet answered, is a difficult task.[27]

As a cancer researcher, I observed much the same psychological struggle, prior to the 1970s, over whether or not to tell a patient they had cancer. Some physicians would not even tell the family of the patient. It was thought that knowing the seriousness of the illness would prevent the patient from any attempt at recovery. They would simply give up hope. However, while it is not easy for either the patient or the family, honesty was always found to be the best policy.

Understanding the value of truth must become part of public policy on the environment. Too often information is suppressed with the excuse that 'people might panic'. We must press for more constructive dialogue and forums in which local community groups can hear clear and honest explanations of any findings. Mutual respect and cooperation will forge a new level of trust between the people and their government.

Health Risks to Children

The full impact of environmental contaminants taken singly or in combination is unknown. Because the pollution build-up is gradual, rates of ill health in exposed adults also escalate over time. Children are especially at risk as they arrive in this world ill-equipped for the elevated levels of toxins we find today. Children differ from adults in their ability to metabolise, detoxify, and excrete toxic chemicals. However, during infancy and childhood bone growth means a higher turnover of bone tissue, so it is sometimes possible to reduce the incorporation of heavy metals including radionuclides – for example, through the use of distilled water.

Environmental health risks to children are increasingly being acknowledged as a key global concern. The United Nations Convention on the Rights of the Child states that a child has the right to enjoy the highest attainable standard of health and health-care facilities and is entitled to a safe environment. This link between the environment and children's health was again recognised on 5 March 1997, when the heads of the United Nations Environment Program and the United Nations Children's Fund signed a Memorandum of Understanding aimed at cooperation in areas fundamental to sustainable development. They will mutually support the implementation of programmes to ensure the well-being of children. If the environment is safe for children, adults are more likely to reach their full potential.

The International Network on Children's Health, Environment and Safety (INCHES) was formed at the meeting of the International Paediatric Society in Amsterdam, August 1998.[28]

INCHES will function as a coordinating structure in which those working on children's health can share the latest data, link with colleagues from other regions, and strengthen the interdisciplinary and international nature of their work. Members will include national and international professional associations, research and policy institutes, advocacy organisations, universities, parent's and children's organisations, national and intergovernmental agencies, and individuals.[29]

The 'State of the World's Children' Report in 1995, which was prepared by Unicef, called attention to the 'increasingly frequent series of catastrophes for children'. The catastrophes included conflicts in Rwanda, Mozambique, Angola, Somalia, the Sudan, Afghanistan, Cambodia, Haiti, and Bosnia. According to Unicef: 'All of these conflicts, made the more devastating by weapons exported from the industrialized nations, brought not only short term suffering to millions of families, but long term consequences for the development of people and of nations'.[30] In the last ten years two million children have been killed in war, four to five million have been physically disabled, more than five million have been forced into refugee camps, and more than twelve million have been left homeless.

The Suffering of Veterans

Environmental health problems generated by military activities pose a unique set of difficulties. In the past, those who tried to link illness to the use of weapons were labelled 'communists' in the West and 'capitalists' in the East. I remember one underground nuclear test in 1970, called the Baneberry event, which went very wrong. The radioactive gas and debris broke through the surface of the Nevada desert and the toxic cloud created travelled across the United States, crossing into Canada near Buffalo, New York. One of the patriotic workers at the test site loudly proclaimed his belief that a 'communist' had done this to discredit the US nuclear testing programme. Most military projects carried this same burden of patriotism.

The toxic nature of military weapons, especially those introduced during and after the Second World War, has caused widespread illness among members of the armed forces. Atomic veterans have formed support organisations in the West and in Russia. These men were marched into ground zero after an atmospheric nuclear explosion to see if they could respond to military commands in the midst of a nuclear war. Yet for these veterans to talk about their experience or even complain about their ill health, which included teeth and hair falling out, severely depressed immune systems and ultimately fatal cancers, was considered unpatriotic.[31] After many years of silence, struggle and rejection, and the deaths of many veterans, some survivors began to receive compensation in the 1990s. In April 2000, the US government finally acknowledged the work-related illnesses of nuclear weapons production workers and announced that they too would be compensated.[32] This is an important first step, since the US has been assuring the rest of the world that radiation protection standards will protect workers from harm.[33]

For years the US government has been operating weapons production facilities under a cloak of secrecy. All claims of ill health were met with court fights in which the burden of proof was on the victim and all of the research, money and legal expertise were on the side of the government. The new compensation programme is being called 'unprecedented' and 'the first tangible acknowledgment of responsibility for decades of unsafe working conditions in dozens of nuclear bomb factories'.[34] It will have ripples for the nuclear industry around the world.

Vietnam war veterans who suffered from exposure to Agent Orange have also formed associations to claim compensation for their illness. Most recently, it is the Gulf War veterans who are sick and dying. These veterans rarely condemn war itself. They are only asking that their government recognise their wounds, extend assistance, and provide care for their families. Perhaps in time these gains will extend to the indigenous people on whose land the uranium for the atomic bomb was mined and milled, and who have been subjected to the polluting effects of weapons testing. Care and assistance is also required for so-called 'downwinders', people living downwind of nuclear test sites and hazardous waste dumps.

A SOUND ECONOMY

Between 1960 and 1993, the death rate for newborns decreased from 28 to 5 per thousand in Finland, while in Niger it remained at 320 per thousand. In all of sub-Saharan Africa the average death rate of newborns was 179 per thousand in 1993, whilst in industrialised countries it averaged 10 per thousand. Much of the disparity was due to inadequate access to safe drinking water, poor sanitation and a lack of health services. In the 'least economically developed' countries only 46 per cent of rural people have access to clean water and only 27 per cent have access to adequate sanitation. Maternal death rates tell an equally grim story. In developed nations about 10 women die per 100,000 births; in developing countries that figure is 351, and in the least developed, it is 607. The relationship between poverty, environmental degradation, health and reproduction is startling.[35]

It is well recognised that building a society based on sustainability requires effective social programmes. In many countries, whether developed or developing, social programmes are under severe strain due to lack of money. Unemployment is entrenched, reducing family income at a time when many government and charity safety nets are either disappearing or being frozen at present levels. This can lead to a sense of desperation and recourse to violence, further undermining the natural world. While everyone agrees that we, as a society, must begin to live within our ecological means, how to achieve this in a just and equitable way is being hotly debated.

A global reduction in the need for military arms would indirectly relieve the debt burden of developing countries, since many contribute a large proportion of their gross domestic product to arms. In fact there have been numerous calls for 'third world debt' to be cancelled altogether so that countries can use their funds internally. Klaus Toepfer, executive director of the UN Environment Program (UNEP), told a press conference, 'Debt relief is a necessary pre-condition for sustainable development – an important step towards increasing resources for education, health and the environment.' He also stated that the relief of debt would only be effective if this were accompanied by good governance: 'Debt relief will only alleviate the living conditions of the billion people who are homeless or live without adequate shelter in our cities if Governments and local authorities commit themselves to better urban governance'.[36]

Some positive moves towards debt relief have been made – in June 1999, world leaders at the G8 summit in Cologne agree to relaunch the Highly Indebted Poor Countries Initiative, and it is expected that some 36 countries will benefit from this, receiving debt write-offs totalling around $70 billion. However, debt relief alone is not going to solve the enormous problems faced by the developing countries of the world, and Klaus Toepfer also called for an increase in overseas development assistance.

The gap between the rich and the poor is widening across the globe. Even in economically developed countries, the mood is quite negative and this is often with good reason. The gross domestic product (GDP) measures the total amount of income-generating production in a society, whether or not that income actually goes to the residents. There is a tremendous difference in the yearly rate of change between the top eight countries and the bottom eight countries.[37] The most recent data available is for 1994, before the crash in the Asian economy.

Top GDP Growth Rate Countries:		Bottom GDP Growth Rate Countries:	
Brazil	10.5%	Canada	0.7%
China	10.3%	U.S.A.	0.5%
South Korea	9.9%	Japan	0.3%
Malaysia	8.9%	Kenya	- 0.2%
Thailand	8.5%	Turkey	- 1.5%
Vietnam	8.5%	Saudi Arabia	- 2.0%
Singapore	8.1%	Russia	- 6.0%
Laos	8.0%	Mexico	-10.5%

On the surface it would appear that the economies of some developing countries such as Vietnam are booming. However, these statistics indicate countries where investment is being made and jobs generated, not necessarily areas where wealth is accumulated. Owners of production may be Canadian, American, or European and therefore the profits go overseas. The Asian crash showed just how volatile these economic gains can be. If external investors recall their input, the entire economy can collapse.

Clearly Canada, the US and Mexico are not areas of job growth, although individuals and transnational corporations within these countries may be producing jobs and gaining large profits elsewhere in the world. For the ordinary citizen in such countries, the experience will be decreased opportunities for work and increased financial worries. At the present time, the force driving the global economic structure is greed, and military power protects that greed.

The good news is that environmental needs directly provide jobs. For example, in Canada alone there are currently more than 5000 environmentally related businesses, employing approximately 300,000 people. Annual sales of this sector are now worth about $22 billion.[38] By strengthening this sector, and forming partnerships between governments, universities and environmental NGOs, perhaps we can begin to build more solid economies that contribute to ecological security rather than destroying it. We can also focus on encouraging industries towards 'clean production'. If industry changes to clean production then worries about environmental health will diminish. As stated in the previous chapter, industry should be creating products that meet a genuine global need rather than luxuries for the wealthy. All of this requires a huge change in the way we think. Teamwork, cooperation and accountability must be the new business paradigm, not competition. It may take time to develop these skills, but it is surely worth the effort.

SOME ENCOURAGING EXAMPLES OF CIVILIAN ACTION

Even if all of this seems difficult to achieve, we should not be discouraged from attempting to bring about more equitable practices. Outstanding individuals have often affected public policy, with or without the backing of a well-known organisation. Working within a coalition is, of course, a powerful way to effect change, and the results achieved by non-governmental organisations throughout the world testify to the strength and potential of citizen action.

Originally, most NGOs were service-giving organisations, providing education in human rights and conflict resolution or assistance to refugees and internally displaced people. These organisations possessed skills, knowledge or influence, which they placed at the disposal of their beneficiaries, hopefully leaving them better off for the intervention. Organisations such as the Red Cross/Red Crescent, Legal Aid, and Doctors Without Borders come to mind. While service was commendable, and unfortunately is still a need, the last decade of the twentieth century saw a broadening of NGO activities to include such areas as political activism and policy formation. These NGO initiatives cause structural changes which will have lasting consequences. A few examples of NGOs are given here, but there are very many more in existence in every part of the world. I hope that the reader will feel inspired by these examples of positive action.

Council on Economic Priorities (CEP)

This is an independent public service organisation dedicated to the analysis of policy on national security, energy and environment, and corporate responsibility. Under corporate responsibility, it analyses ethical investments, political action committees, fair employment and consumer issues. In 1986 CEP issued its rating of 'America's Corporate Conscience' and a consumers' 'Shopping for a Better World'. In 1990 it began issuing its Corporate Conscience Awards, and is now working globally to certify corporations as socially responsible.[39]

International Society for Ecological Economics (ISEE)

This provides a major forum for economists, ecologists, academics and activists to link issues and devise strategies. One of the key figures in its formation was Herman Daly, an economist with experience in the United States, Brazil and Australia, who has written extensively on the relationship between economics, ecological sustainability and ethical behaviour. This organisation makes a major contribution to the understanding of how economics contribute to environmental destruction, and also gives viable ideas about what can be done in the future.[40]

Committee of Soldiers' Mothers (Russia)

Committee of Soldiers' Mothers was formed by Russian women in 1989, originally to bring their sons home from military service so that the young men could go to school. They have succeeded in bringing home 180,000 young men for this purpose. The Mothers have also protested against many of the behaviours they discovered were common practice in the military, such as regular beatings, abuse and humiliation. They have found that the military lacked sufficient food and other necessities, and that about 30 per cent of soldiers were being used for construction and were virtually treated like slaves. Some of the Mothers' demands for reform and for civilian oversight of the military were conceded by President Gorbachev, but most were not respected. The Mothers then set up a rehabilitation centre for soldiers who left the military for health reasons, and this was expanded to include human rights education for conscripts and their parents. The Mothers worked on legislative proposals and non-violent protests. During the Chechnya War, hundreds of Mothers went to Chechnya to bring their sons home. They organised a 'March of Mother's Compassion' and bombarded the Russian Duma with petitions and statements. The organisation was started by just five women.[41]

Science Writers' Forum of Kerala, India (KSSP)

This organisation originally focused on communication of science in the local vernacular. It became a People's Science Movement, inspiring many similar organisations to spring up in India. It now has 60,000 members organised in about 2000 units. It provides in-service teacher training, assesses curricula and textbooks, promotes innovation, publishes science books and journals for children and runs massive Children's Science Festivals and teacher exchange programmes. Through the group's assistance, Kerala achieved total literacy by 1991. The KSSP has had substantial influence on health, women's issues, research and development. Most of its income comes from its publications, including the selling of books door to door, and it has received no foreign financial assistance.[42]

Sahabat Alam Malaysia-Sarawak (SAM)

Friends of the Earth in Malaysia has been involved in campaigns on environmental health, indigenous rights, protection of tropical forests, logging, pollution, soil erosion and land spoilage. In spite of a sometimes repressive government, SAM, together with other influential Malaysian organisations such as the Consumer's Association of Penang (CAP), Asian-Pacific People's Environment Network and Third World News Network, has focused civil society and government on major problems of inequitable development policies, loss of indigenous seeds, pesticide contamination, corporate responsibility and sustainable agriculture. They pioneered the concept of environmental reporting with their State of the Malaysian Environment Report of 1983–84.[43]

Centre for Development Alternatives

Set up in Chile in 1981 by Manfred Max-Neef, the Centre for Development Alternatives aims to practise 'economics as if people mattered'. The organisation attempts to reorient development so that it stimulates self-reliance and satisfies basic human needs. It is a clearinghouse for information on revitalisation and development of small and medium-sized communities, both rural and urban. Manfred Max-Neef is doing seminal work on human needs (having, doing and being) and ethical values.[44]

The Future in Our Hands

This Norwegian institute researches and reports regularly on alternative political solutions to crises. It promotes the need for a society which puts 'social, global and environmental values above economic considerations' and coordinates corresponding movements both in the industrial and the developing world. It has an annual budget of $3 million, which it uses to fund projects in more than 20 countries.[45]

MOVING FORWARD

Over the last 50 years, the various agencies of the United Nations have evolved rather miraculously into more or less effective agents for global change. This is a remarkable accomplishment made difficult by the multinational/multilingual nature of its staff and the escalating nature of its mandate. The world's expectations of it are awesome, and it has endured great financial insecurity. At this crucial time in history, it is important to re-think the structure of the UN, its agencies and their mandates so as to encourage further growth in the direction of sustainability and genuine security. It is also important that human security be redefined as the new vision towards which we strive.

Because the UN lacked physical coercive power from the beginning, it has developed the more feminine qualities of consensus building and moral persuasion. Recently, through a quick succession of international conferences, the UN has been building support for an agenda that promotes environmental protection, sustainable development, human rights, population control, and the rights of women and children. While this consensus has not yet been translated into action, I believe that its vitality will prevail.

Women have often been the agents of social change. In recent times we have had two women heads of leading UN agencies, Elizabeth Dowdeswell for UNEP and Carol Bellamy for Unicef. Barbara Ward co-authored the book *Only One Earth* with Rene Dubois. This book was vitally important in putting the environment on the agenda for the 1972 United Nations Conference in Stockholm. In particular the book helped to bring a great deal of legitimacy to governmental concern for the social, political and economic dimensions of environmental change. The World Women's Conference in Beijing in 1995 was incredibly well attended and for the first time in history, the NGO parallel conference was better attended than the governmental conference. These are all encouraging signs.

However, powerful nations and multinational corporations have been trying to orientate the UN towards the needs of global trade in an aggressive market economy. The theory is that this will produce jobs and an increased standard of living for everyone. The reality is otherwise. The glorification of commerce is linear thinking at its worst and betrays little concern for either the environment or for social welfare. As the Western world moves from military competition to exerting firm control over world trade policies, it has been attempting to use the UN to consolidate its control. The outcome of this struggle for power and influence within the UN will have enormous implications for the future health of the planet.

It seems to me that the present global trade war offers the same unsustainable scenario as a shooting match. It imposes the 'logic' of structural adjustment within struggling nations with developing economies, which in turn brings to the people poverty, sickness and crime. The 'cure' creates the problem. Such instability, with the poor growing poorer and the rich growing richer, cannot be 'managed' with brute force or suppression. Greed, violence and short-term goals will mean a 'good life' for a few people for a short time but will result in the ultimate destruction of the environment and society as we know it. The alternative is widespread behavioural change and the adoption on a grand scale of attitudes, values and behaviours that lead to sustainable development. It has been slow in coming, but when one believes that life is stronger than death, one can stand against the tide and keep waiting and working for this change.

When the UN was first launched in 1945, there was a vision born with it to nourish its more practical structural components. This vision was embodied in the Human Rights Covenants. They applied to every person born into this world and have given rise to many organisations, such as Amnesty International and the International Human Rights Lawyers. The covenants have given pause to those who would abuse power, and even where they failed to moderate aberrant behaviour, they took away all pretence to social approval. Humans *can* change their behaviour even when that behaviour was not questioned in the past. We no longer view slavery, torture, oppression of women, exploitation of children and destruction of worker health as acceptable. The struggle goes on to outlaw capital punishment, genocide, rape, and violence.

In our day, we have another new vision that can grow over the next decades. This vision is embodied in the Earth Charter. If supported, it will not only broaden the concept of good global governance and citizenship, but it should ultimately ensure the security of our planet.

CONCLUSION

The anti-nuclear peace movement is now 50 years old and the environmental movement 25 years old. I have never heard a good analysis of environmental problems caused by military research and development, or a good analysis of the environment used as a military weapon. Perhaps it is because the peace movement looks more at war itself than the preparation for war. Most of the efforts of the peace movement have gone into managing and reducing the risks of nuclear war, not into careful monitoring of more general military research and its impact on Earth. Older peace movements have been occupied with low-intensity warfare, ethnic hatred, genocide, and various other horrors. These are all important issues. However, a simple strategic move like cutting off the money and personnel for military research would, I think, effectively stop the escalation of violence.

The environment movement, on the other hand, focuses primarily on the impact of civil society, lifestyles and multinational corporations with very little analysis of the colossal impact of war. It is no wonder that 25 years of environmental education and effort has not reversed even one major environmental problem. The ozone layer is disappearing at a rate faster than predicted; deforestation and desertification are widespread; ten thousand species are lost each year; both infectious and chronic disease rates are increasing globally; toxic waste is piling up, being inappropriately isolated, even being shipped to unsuspecting communities in developing countries. Weather and climate change is being blamed on El Niño, but no one explains why El Niño is suddenly so frequent and so extreme. If the observed Earth spasms are not directly connected with atmospheric experimentation, they are certainly evidence of atmospheric instability and of the danger of plans to destabilise it further.

In this book I have tried to give the reader some idea of the current crisis in behaviour which, if continued, will bring the global community to its knees. Recovery will be difficult since both human health and the planet's life support system are being attacked. As new generations are born they must cope with the mistakes of past generations – depleted resources, polluted land, economic and environmental instability. Excessive greed dominates the global economy, promoting deprivation and resentment, which will eventually erupt in violence. Weaponry is potentially more destructive than ever before and targets not just people and buildings, but the very structure of the Earth itself.

On the bright side, there is a growing awareness of the problems and an important infrastructure for healing is slowly being built up, mostly through the actions of dedicated leaders in the global sector and hard-working volunteers in the community. Networking and coalition-building are the way ahead, with the Internet being one of the best tools for global organising. A few of the many thousands of forward-thinking organisations have been mentioned in this book, but there are many more. However, it was not the purpose of this book to cover all of the opportunities open to each and every person to help abolish war, exploitation and environmental destruction. Rather, it has been my aim to show that war needs to be abolished and that the time is ripe for this to happen.

The important message of this book is that local problems and local solutions are of tremendous importance. No positive initiative and fruitful discovery should be lost. Whatever seems to be a local problem needs to become connected to the larger body politic, whether the city, the state or province, nation, region or world. No matter on which level a person works, or how much praise or pay they receive or fail to receive, their efforts form part of the network that is building the future.

I hope this book has given readers some inspiration as to how they might become involved in helping this peaceful planet evolve to its full potential. Despite years of abuse, it is still an amazing and beautiful creation. It deserves our best efforts. Enjoy it, love it, and save it!

Appendix A

ORIGINAL EARTH COUNCIL GOVERNING BOARD

Maurice Strong, Chair, Canada

H.R.H. Princess Basma Bint Talal, Chair of The Queen Alia Fund for Social Development, Jordan

Elizabeth Evatt, member and Former President, Australian Law Reform Commission, Australia

Arnoldo Jose Gabaldon, Former Minister of the Environment, Venezuela

Jose Goldemberg, University of São Paulo, Institute of Electrotechnics and Energy, Brazil

Gordon T. Goodman, Former Chair of the Stockholm Environment Institute, UK

Abdlatif Y. Al-Hamad, Director General and Chair of the Board, Arab Fund for Economic and Social Development, Kuwait

Mahbub ul Haq, President, Human Development Centre, Pakistan

Saburo Kawai, Chair, International Development Center of Japan, Japan

Tommy Koh, Director, Institute for Policy Studies, Singapore

Vladimir Mikhailovich Kotlyakov, Director, Institute of Geography, Russian Academy of Sciences, Russia

Jonathan Lash, President, World Resources Institute, USA

Emile van Lennep, Former Secretary-General, OECD, The Netherlands

Robert Lion, President, Energy 21, France

Wagaki Mwangi, Executive Coordinator, Econews Africa, Kenya

Bisi Ogunleye, National Coordinator, Country Women Association of Nigeria, Nigeria

Ambassador Mohamed Sahnoun, Algeria

Emil Salim, Former Minister of Environment, Indonesia

Klaus Schwab, President, World Economic Forum, Germany

Academician Sun Honglie, Member, the Presidium of Chinese

Academy of Sciences, China
Pauline Tangiora, Women's International League for Peace and Freedom/Maori Women's Welfare League, Rongomaiwahine Tribe, Aotearoa/New Zealand

For more information contact:
Earth Council
Apartado 2323-1002
San Jose, Costa Rica
Tel: +506-256-1611
Fax: +506-255-2197
Email: *eci@terra.ecouncil.ac.cr*
Website: *http://www.ecouncil.ac.cr*

New York office:
Tel: +1-212-682-5998

Benin office:
Tel: +229-314-023

Appendix B

THE INTERNATIONAL COUNCIL FOR LOCAL ENVIRONMENT INITIATIVES

WORLD SECRETARIAT
City Hall, East Tower,
8th floor
Toronto ON M5H 2N2
Canada
Tel: +1-416-392-1462
Fax: +1-416-392-1478
Email: iclei@iclei.org
Web: http://www.iclei.org

EUROPEAN SECRETARIAT
Eschholzstrasse 86
D-79115 Freiburg,
Germany
Tel: +49-761-368920
Fax: +49-761-36260
Email: *100757.3635@compuserve.com*

ASIAN PACIFIC SECRETARIAT
Japan Office
c/o Global Environmental Forum
Iikura Building,
3rd floor
1-9-7 Azabudai, Minato-ku
Tokyo, 106 Japan
Tel: +81-3-5561-9735
Fax: +81-3-5561-9737
Email: *100506.1062@compuserve.com*

OFFICE OF THE AFRICA REGIONAL COORDINATOR
108 Central Avenue,
PO Box 6852
Harare, Zimbabwe
Tel: +263-4-728984
Fax: +263-4-728984
Email: *iclei@zol.co.zw*

OFFICE OF THE LATIN AMERICA REGIONAL
COORDINATOR
I. Municipalidad de Santiago
Corporacion para el Desarrollo de Santiago
Av. Cardenal Jose Maria Caro 390
PO Box 51640 Correo Central
Santiago, Chile
Tel: +56-2-632-9665
Fax: +56-2-638-3112
Email: *icleila@cmet.net*

US OFFICE
15 Shattuck Square, Suite 215
Berkeley, California 94704 USA
Tel: +1-510-540-8843
Fax: +1-510-540-4787
Email: *75463.3516@compuserve.com*

NOTES

Chapter 1

1. 'MPs say Kosovo bombing was illegal but necessary', *Guardian*, 7 June 2000.
2. This is a 46-page US policy statement. Important sections were excerpted in a *New York Times* article published on 8 March 1992.
3. 'US Defense Planning Guide', Pentagon, 1992.
4. *NATO in the Balkans*, International Action Center, 39 West 14th Street, 206, New York, NY 10011 (web site: *www.iacenter.org*), 1998.
5. Emil Vlajki, *The New Totalitarian Society and the Destruction of Yugoslavia*, Legas Press, New York, NY, 1999.
6. For a complete history of the OSCE, see web site *http://www.osce.org*.
7. Rollie Keith, in 'Failure of Democracy', *The Democrat*, May 1999.
8. Sveriges TV/radio interview, Stockholm, 24 May 1999.
9. Johan Galtung, Professor of Peace Studies, Transnational Foundation for Peace and Future Research, Vegagatan 25, S-224 57 Lund, Sweden.
10. Bradley Graham, 'Medals Granted after Acknowledgment of US role in El Salvador', in *Washington Post*, 6 May 1996.
11. 'Les morts de Racak ont-ils vraiment été massacré froidement?', *Le Monde*, 21 January 1999, p2; and 'Kosovo: zones d'ombre sur un massacre', *Le Figaro*, 20 January 1999, p3. See also the excellent analysis of the 'failure' of the OSCE: Diane Johnstone, 'Making the Crime Fit the Punishment', in *Masters of the Universe? NATO's Humanitarian Crusade*, edited by Tariq Ali, Verso, London, 1999.
12. For a post-war understanding of what happened at Racak see 'Yugoslav Government War Crimes in Racak', *Human Rights Watch*, 29 January 1999.
13. Richard Foot and Patrick Graham, 'Australians criticize CARE Canada for political work in Balkans', *National Post*, 3 February 2000.
14. Rollie Keith, in 'Failure of Democracy'.
15. *Depleted Uranium: A Post-war Disaster for Environment and Health,*

Laka Foundation, Ketelhuisplein 43, 1054 RD Amsterdam, The Netherlands, May 1999.

16. Wayne C. Hanson and Felix R. Meira Jr, 'Long-term Ecological Effects of Exposure to Uranium', Los Alamos Scientific Laboratory, July 1976.

17. Richard L. Fliszar, Edward W. Wilsey and Ernest W. Bloore, 'Radiological Contamination for Impacted Abrams Heavy Armour', Report No. BRL-TR 3068, December 1989.

18. UK statutory dose limits are set by the National Radiation Protection Board and were quoted in the annual reports of the Atomic Weapons establishment, Aldermaston.

19. UN Secretary General's Report, Document No.E/CN.4/Sub.2/ 1997/27, issued 24 June 1997.

20. Felicity Arbuthnot, 'Depleted Uranium Warning Only Issued to MoD Staff', Sunday Herald, August 1 1999.

21. The Boston Globe, 6 August 1999, A1. The quote is from Pekka Haavisto, the head of one of the UN environmental teams.

22. UN News Release No. 70, 1999.

23. The map of DU ammunition use in Kosovo is available at: http://www.antenna.nl/wise/uranium/img/dukosom.gif.

24. International Court of Justice press communique 99/23, dated 2 June 1999.

25. Anthony Goodman, Reuters News Agency, 'No War Crimes Probe into NATO Bombing', The Toronto Star, 3 June 2000.

26. Testimony of Ramsey Clark, former Attorney General, May 1991, at the Commission of Inquiry for the International War Crimes Tribunal, New York.

27. Testimony in 1989 by CIA Director William Webster before the US Congress.

28. William Thomas, Bringing the War Home, Earthpulse Press, Anchorage, AK, 1998, Ref. no. 29, pp16–17.

29. Testimony of Ramsey Clark, op cit.

30. The Riegle Report to the US Senate Committee on Banking, Housing and Urban Affairs, 25 May 1994; see also William Thomas, Bringing the War Home, Appendix VII, p428.

31. Stockholm International Peace Research Institute ARMS Transfer Project.

32. William Thomas, op cit, p19.

33. 'Ex-US Envoy Misled US on Saddam, Baker told', Toronto Star, 12 July 1991, A3, Washington (Special) reprinted from the Los Angeles Times.

34. John Pilger, 'Mythmakers of the Gulf War', *Guardian Weekly*, 13 January 1991.
35. Testimony of Ramsey Clark, op cit.
36. Ibid.
37. P. Young and P. Jesser, *The Media and the Military*, St Martin's Press, New York, NY, 1997, p 2.
38. 'Economies with Truth', *Guardian Weekly*, 13 January 1991.
39. Thierry D'Athis and Jean-Paul Croize, *Golfe: la guerre cachee*, Jean Picollec, Paris, 1991.
40. Op-ed column by Ramsey Clarke in the *Toronto Star*, 18 February 1991.
41. Ibid.
42. From a briefing document prepared for Canadian members of parliament, Ottawa, 1 February 1991.
43. Norman Friedman, 'Desert Victory: the War for Kuwait', US Naval Institute, Annapolis, 1991.
44. Reported in *The Times* and quoted in William Thomas, op cit.
45. The Hersh article 'Overwhelming Force: Annals of War' appeared in the 22 May 2000 issue of *The New Yorker* magazine and was discussed by Michael R. Gordon in 'Report Revives Criticism of General's Attack on Iraqis in '91', *New York Times*, 15 May 2000.
46. William Thomas, op cit, p101.
47. Testimony of Ramsey Clark, op cit.
48. 'Depleted Uranium – deadly weapon, deadly legacy?', Nick Cohen, *Guardian*, 9 May 1999.
49. Dan Fahey, 'Case Narrative: Depleted Uranium Exposures', Swords to Plowshare, National Gulf War Resource Center and Military Toxics Project; latest update 2 March 1998.
50. 'Health and Environmental Consequences of Depleted Uranium Use in the US Army', US Army Environmental Policy Institute (AEPI), June 1995 pp 83–4.
51. William Thomas, op cit, p97.
52. Robert Fisk, 'Iraqi Casualties Remain a Mystery', *The Independent News Service*, in the *Toronto Star*, 5 August 1991, A5.
53. Rich McCutcheon, 'From Baghdad to Karbala: The Human Cost of War', April 1991, informally distributed to friends and concerned citizens.
54. Ann Montgomery, 'Iraq: The Suffering Continues', in *Ground Zero*, Winter 1991–1992.
55. *Nurses for Social Responsibility*, Vol. 6 No. 2, Summer 1991.
56. Bruce McLeod, 'From Canada's Heart', *The Toronto Star*, 3 January 1992, A17.

57. *The Washington Report on Middle East Affairs*, July/August 1995, p. 105.
58. 'Children in Iraq Suffering, UNICEF', *The Toronto Star*, 27 November 1997, A21.
59. 'Death Rates Rising in Iraqi Children', *Reuters News*, 26 May 2000.
60. This is based on satellite images and not the Pentagon propaganda which had exaggerated the amount of the spill. *The Daily News*, Halifax, 22 February 1991, p8, from 'Truth Behind Gulf Slick May Be a Victim of War' by Keay Davidson, *San Francisco Examiner.*
61. According to the brief drawn up by Ramsey Clark, US aircraft caused much of the worst oil spills. Allied helicopters dropping napalm and fuel air explosives on oil-well storage facilities, storage tanks and refineries caused oil fires throughout Iraq and many, if not most, of the oil-well fires in Kuwait. Those not started by the US were deliberately caused by the fleeing Iraqi military to help cover their escape.
62. British biologist J.L. Cloudsley-Thompson, quoted in a Parliamentary Briefing prepared for Jim Fulton, MP, 22 January 1991.
63. Steve Newman, 'Earth Week: a Diary of the Planet', *The Toronto Star*, D6, Saturday 10 August 1991.
64. John Horgan, 'Burning Questions: Scientists Launch Studies of Kuwait's Oil Fires', *Scientific American*, July 1991, pp17–24.
65. Ibid.
66. Ibid, p20.
67. Ibid, p17.
68. Keith Schneider, 'Environmental Rule is Waived for Pentagon,' *The Times*, 30 January 1991.
69. 'Bombing Strikes Stepped up in "Secret War" against Iraq', *Guardian*, 8 June 2000.
70. See, for example, 'Under UN Cover', Editorial, *Washington Post*, 3 May 1999 and Thomas W. Lippman and Barton Gellman, 'US Said it Collected Iraq Intelligence via UNSCOM', *Washington Post*, 8 January 1999.
71. John Pilger, *Paying the Price: Killing the Children of Iraq*, Carlton, 2000.
72. US newswire, 9 June 2000.
73. Statement of 'Michael Rackman' to the Subcommittee on Resources and Intergovernmental Relations House Committee on Governmental Reform and Oversight, 26 June 1997. The name has been changed to protect the identity of the veteran.

74. Protocol I Additional to the Geneva Convention – 1977; Part I, Chapter III, Article 54.

75. Personal correspondence from participant Larisa Skuratovskaya, 26 May 2000.

76. In 'Metal of Dishonor', International Action Center, New York, NY, 1997, pp172–73.

Chapter 2

1. There is also some life in the deep oceans that depends on Earth energy rather than solar energy. This life forms near rifts on the ocean floor, where there are hydrothermal vents. These vents provide a warm spot in the deep ocean, and spew out mineral nourishment for living organisms such as giant clams and tube worms 3.7 metres or 12 feet long.

2. An orbit is the path followed by an object when it stops deploying its artificial power source and allows its motion to be governed solely by the force of gravity. It is sometimes called 'free fall'. When you throw a ball, the height it reaches depends on your thrust power. At some point the ball runs out of this 'thrust' and begins to fall back to the Earth in a curved path – it does not stop abruptly and fall straight back to Earth but continues in a forward direction as it descends. If you reach a sufficient height above the Earth – for example, by using a rocket – the path of the falling object will match the curve of the Earth.

3. Grolier *Multimedia Encyclopaedia*, 1996 and 1998, and Microsoft *Encarta Multimedia Encyclopaedia*, 1999, are good sources of information on the atmosphere and the Van Allen belts. See also B. Hultqvist and C.G. Falthammer (eds), *Magnetospheric Physics*, Kluwer Academic Publishers, 1990, for greater detail.

4. Gary V. Latham, 'Moon', 1998, in Microsoft *Encarta Multimedia Encyclopaedia*.

5. George P. Sutton, 'Rockets and Missiles', 1999, in Grolier *Multimedia Encyclopaedia*.

6. For more on nuclear weapons testing, see Rosalie Bertell, *No Immediate Danger: Prognosis for a Radioactive Earth*, The Women's Press, London, 1985.

7. *New York Times*, 19 March 1959.

8. *Globe and Mail*, 1 December 1988.

9. C.E. Miller and L.D. Marinelli, 'Measurement of Gamma Rays

Activities from the Human Body', Argonne National Laboratory Report No. 5518, 1956.

10. Linden Kurt, 'Cesium 137 Burdens in Swedish Laplanders and Reindeer', *Acta Radiologica*, Vol. 56, p237, 1961.

11. Unpublished studies by Peter M. Bird, PhD, Environment Canada, Government of Canada, Ottawa, May 1965.

12. 'Canada and the Human Environment, English Summary,' paragraph 3.11, Government of Canada, June 1972.

13. 'Incidence of Neoplastic Diseases in Canadian Eskimos' [a name no longer used for the Inuit], a letter to the Editor, *Canadian Medical Association Journal*, Vol. 82, 30 January 1960, pp280–281.

14. See Bertell, *No Immediate Danger*, op cit, pp20–63.

15. 'The changing picture of neoplastic disease in the Western and Central Arctic 1950-1980', *Canadian Medical Association Journal*, Vol. 130, 1 January 1984.

16. 'Soviet herders suffer effects of nuclear tests', Associated Press, Moscow, quoted in *Japan Times*, 18 August 1989.

17. Keesings Historisch Archief (K.H.A.), 13–20 August 1961.

18. Nigel Harle, 'Vandalizing the Van Allen belts', *Earth Island Journal*, Winter 1988–89, p11.

19. Nick Begich and Jeane Manning, *Angels Don't Play This HAARP*, Earthpulse Press, Anchorage, AK, 1995, p53.

20. K.H.A., 29 June 1962.

21. K.H.A., 11 May 1962.

22. Christmas Island is located at 2 degrees north latitude and 157 degrees west longitude. In spite of its use for nuclear testing and the residue of radioactivity, Christmas Island is now again 'open for business' and being promoted for tourism and other industries.

23. K.H.A., 11 May 1962.

24. K.H.A., 5 August 1962.

25. As explained in 'The High Energy Weapons Archive', in Science & Technology, *Encyclopaedia Britannica*, 2000.

26. *Long-term Effects of Multiple Nuclear-weapon Detonations*, US National Academy of Science, 1975.

27. M. Mendillo, et al., *Science*, Vol. 187, p343, 1975.

28. *Environment News Service Daily*, 13 February 1992.

29. Grolier *Multimedia Encyclopaedia*, August 1996.

30. *Long-term Effects of Multiple Nuclear Weapons Detonations*, op cit.

31. 'Istanbul, Turkey at the Habitat II Summit,' Reuters, 4 June 1996. Figures are according to the World Bank.

32. *USA Today*, 21 July 1994, p3A.
33. Committee to Bridge the Gap, letter to Dr Dudley McConnell, deputy director, Advanced Programs, Solar System Exploration Division (Code EL, NASA, 6 April 1990).
34. Speech given outside the gates of the Cape Canaveral Air Force Station, 24 June, 1997, at a Florida Coalition for Peace and Justice demonstration against Cassini.
35. Robyn Suriano, 'NASA Worker Suspended for 2 Days', *Florida Today*, 13 September 1997.
36. 'Space Accidents of the Past', Associated Press, 25 June 1997.
37. Professor Ernest Sternglass, personal communication, 7 August 2000.
38. *Aviation Week & Space Technology*, 5 August 1996. General Ashy is also commander of the US Air Force Space Command and commander-in-chief of the combined US–Canada North American Air Defense Command (NORAD).

Chapter 3

1. The Grolier *Multimedia Encyclopaedia*, 1996 citing Charles Sheldon et al., *Soviet Space Programs*, US Government publication, 1971 (revised 1976 and 1981).
2. Citizen Energy Project Study Brief on the Solar Power Satellites (SPS), 1978. This and other briefs were submitted to the US government during the comment period on the SPS project and no doubt exist somewhere in the US national archives. The set kept by Rosalie Bertell is in the Bertell Archive, Canadian National Archives, Ottawa, Canada.
3. Grolier *Multimedia Encyclopaedia*.
4. Extra-low frequency is the wavelength needed to reach submerged submarines by radio.
5. 'Solar power white paper on military implications', critique of SPS by Michael J. Ozeroff, SA-1, 1978, part of the Citizen Energy Project Study Brief on the Solar Power Satellites (SPS), (see ref 2 above).
6. *Popular Science*, September 1997.
7. Televised statement, later quoted by the *Globe and Mail*, 11 March 1991.
8. William Saphire, 'The Great Missile Mystery', *New York Times*, 11 March 1991, A1.
9. William E. Burrows, in Grolier *Multimedia Encyclopaedia*.
10. This explanation is based on information in Richard Wolfson and Jay

M. Pasachoff, *Physics with Modern Physics: For Scientists and Engineers,* second edition, HarperCollins College Publishers, New York, NY, 1995.

11. The total US budget for the fiscal year 1996 was $41.6 trillion, $256 billion of which went to defence.
12. *Jane's Defence Weekly,* 25 February 1989.
13. Rich Garcia, 'Airborne laser arrives in Wichita', Air Force Research Laboratory Public Affairs, *Air Force News,* 24 January 2000.
14. Courtesy of Air Force Material Command News Service, 24 January 2000.
15. Associated Press, Washington, 3 September 1997.
16. Alert from Global Network Against Weapons and Nuclear Power in Space, Florida, Fall 1999.
17. 'Clinton Lawyers Give Go Ahead to Missile Shield', *Washington Post,* 15 June 2000.
18. 'More Doubts Raised on Missile Shield', *Washington Post,* 18 June 2000.
19. 'GAO Report Finds Fault with Missile Shield Plan', *Washington Post,* 17 June 2000.
20. Harry Mason, 'Bright Skies part I', *Nexus,* March–April 1997.
21. Harry Mason, 'The Banjawarn "Bang" Revisited', *Nexus,* June 1997.
22. Harry Mason, 'Bright Skies part I', op cit.
23. Encarta *Multimedia Encylopaedia,* 1999.
24. 'Tremorous night of the death ray', *New Zealand Herald,* 25 January 1997.
25. David Shukman, *Tomorrow's Wars: The Threat of High-Technology Weapons,* Harcourt Brace & Co., New York, 1996, p174.
26. Dr Huda S. Ammash, 'Toxic Pollution, The Gulf War, and Sanction', in *Iraq Under Siege: The Deadly Impact of Sanctions and War,* Anthony Arnove (ed), South End Press, Cambridge, MA, 2000.
27. Catalogued in the 1981 US Department of Transportation's 'First Annual Workshop on Aviation Related Electricity Hazards'.
28. Patricia Axelrod, 'Disaster Signals: Suicide Weapons', *International Perspectives in Public Health,* Vol. 6, 1990, pp10–20.
29. Technical Investigation BB61 Addendum 3-Status as of June 1989.
30. Patricia Axelrod, op cit.
31. 'Worldwide US Military Active Duty Military Personnel Casualties, October 1, 1979, through September 20, 1998', US Department of Defense Directorate for Information and Reports Booklet M07, 1980.

32. 'Nuclear Disarray', Bruce Nelan, *Time*, 19 May 1997.
33. 'Pentagon Envisions Cyber-warfare Rise', *Washington Times*, 31 May 2000.
34. National Aeronautics and Space Administration (NASA) Lyndon B. Johnson Space Center in Houston, Texas, January 1987.
35. See, for example, the *guardian–unlimited.co.uk* website or *http://www.networkingusa.org/fingerprint/page1/fp-political-control.htm*.
36. Yorkshire CND
37. Steve Wright, 'Assessing the Technologies of Political Control', Omega Foundation, Science and Technology Options Assessment (STOA), Dick Holdsworth (ed), 6 January 1998. The 112-page STOA report is available from the European Parliament. See also 'Britain and US Accused in Spy Row', *Guardian*, 5 July 2000.
38. Steve Wright, op cit.
39. Ben Barber, 'Group will battle propaganda abroad', *Washington Times*, 28 July 1999.
40. Information based on extensive confirmation obtained by Geoff Metcalf, staff reporter for *WorldNet Daily*.
41. Office of Research and Engineering, US National Transportation Safety Board report, by Vernon S. Ellingtad, of 20 October 1998. No cause of the destruction of TWA flight 800 has been determined and apparently the investigation is still open, as officially declared at the conclusion of the public hearings on 12 December 1997, by Alfred W. Dickerson, investigator in charge.
42. James Sanders, *The Downing of TWA Flight 800: The Shocking Truth Behind the Worst Airplane Disaster in US History*, Zebra Books, Kensington Publishing Corp., New York, NY, 1997.
43. AEGIS is a radar and target management system used by the US navy.
44. James Sanders, op cit, pp23–24.

Chapter 4

1. E.L. Heacock, 'Remote Sensing and Meteorology: A Review of the State of the Technology and Its Implications', in *Outer Space: A Source of Conflict or Cooperation*, Bhupendra Jasani (ed), United Nations University Press in cooperation with the Stockholm International Peace Research Institute (SIPRI), Tokyo, 1991, pp69–90.
2. 'Space Programs, National', Grolier *Multimedia Encyclopedia*, August 1996 (many more references are given in the encyclopedia for more detailed research on the global space industries).

3. The International Atomic Energy Agency (IAEA) promotes nuclear power even in developing countries and has claimed primacy over the World Health Organization (WHO) in interpreting the health effects of nuclear radiation, based on a Memo of Understanding with WHO in 1959. It is because of this agreement, for example, that IAEA was the primary agency interpreting the contamination and damage to human health arising from the tragedy at Chernobyl.

4. W.N Hess, *Weather and Climate Modification*, Wiley, New York, 1974.

5. This was covered in US Congressional subcommittee hearings on Oceans and International Environment in the 1970s.

6. Zbigniew Brezinski, *Between Two Ages: America's Role in the Technetronic Era*, Penguin Books, Cambridge, MA, 1976.

7. Lowell Ponte, *The Cooling*, Prentice-Hall, Inc., Upper Saddle River, New Jersey, 1976.

8. Michael Rycroft, 'Active experiments in space plasmas', *Nature*, vol. 287, 4 Sept. 1980, p7.

9. C.N.263.1978.Treaties-12 (Convention on the Prohibition of Military or Other Hostile Use of Environmental Modification Techniques, 10 December 1976, UN General Assembly).

10. See *United Nations Registry of Space Objects*, compiled by Jonathan McDowell, Harvard University, Cambridge, MA, 1997.

11. 'Night Clouds Won't Have Silver Lining but Will Be Red, Blue, Scientists Say', *Buffalo News*, 10 January 1991.

12. 'Northern Lights Thrill Sky Watchers from Texas to Ohio', *Kansas City Star*, 10 November 1991.

13. Information obtained from the Parliamentary Library, Ottawa, Canada at the request of Jim Fulton, Member of Parliament.

14. Richard Wolfson and Jay M. Pasachoff, *Physics with Modern Physics*, second edition, HarperCollins College Publishers, New York, NY, 1995.

15. More information on HIPAS can be found at: *http://www.hipas.alaska.edu*.

16. 'HAARP: HF Active Auroral Research Program', Joint Service Program Plans and Activities: Air Force Geophysics Laboratory and Navy Office of Naval Research, February 1990.

17. Nick Begich and Jeane Manning, *Angels Don't Play This HAARP*, Earthpulse Press, Anchorage, AK, 1995.

18. This can be found in a supplement to the National Industrial Security Program Manual, released in draft form in March 1992.

19. David Yarrow quoted in *Angels Don't Play This HAARP*, op cit, p73.
20. ARCO is the US corporation that has the construction contract for HAARP.
21. Bernard Eastlund, 'Applications of in situ generated relativistic electrons in the ionosphere', Eastlund Scientific Enterprises Corp, 13 December 1990.
22. *New York Times*, 22 September 1940.
23. 'C3 systems' means Command, Control and Communication Systems.
24. 'HAARP: HF Active Auroral Research Program', Joint Service Program Plans and Activities: Air Force Geophysics Laboratory and Navy Office of Naval Research, February 1990.
25. Ibid., paragraph 4.1.1.
26. *Angels Don't Play This HAARP*, op cit, p64.
27. The Proceedings of the 1988 International Tesla Symposium, Reno, Nevada.
28. Tracey C. Rembert, 'Discordant HAARP', from 'Currents', *E Magazine, Britannica.com*, 1 January 1997.
29. John Mintz, 'Pentagon flights secret scenario speculation over Alaskan antennas', *Washington Post*, 17 April, 1995.
30. Grant proposal, 'Arctic Research Initiative: expansion of the SuperDARN Radar Network', submitted to Dr. Odile de la Beaujardiere, National Science Foundation, Division of Polar Programs, 5 November 1996, by S.M. Krimigis, Head, Space Department, The Johns Hopkins University Applied Physics Laboratory. Ref. No AC-23434.
31. *http://w3.nrl.navy.mil/projects/harp/faq.html*.
32. US National Science Foundation awarded proposal OPP-9704717, Applied Physics Laboratory of Johns Hopkins University. Obtained under the Freedom of Information Act by Ms Kristin Stahl-Johnson, Kodiak, AK, 21 April 1998.
33. Operator: Applied Physics Laboratory, Johns Hopkins University, Baltimore, MD, as identified by the awarded grant (see above).
34. Sheila Ostrander and Lynn Schroeder, *Super-Memory: The Revolution*, Carol and Graf Publishers, New York, NY, 1991, p299.
35. Human brain waves are generally between 4 and 35 Hertz. Children's brain waves tend to have lower wave frequencies, in the 4–7 Hertz range. Reflective or meditative brain waves in adults are in the 8–12 Hertz range, and alert active adult brain waves are in the 13–35 Hertz range.
36. See a description of Woodpecker in *Specula Magazine*, January 1978.

37. Bill Sweetman, *Aurora: The Pentagon's Secret Hypersonic Skyplane*, Motor Books International, Oscela WI, 1993, pp152–169.
38. See Harold Puthoff, 'Everything or Nothing', *New Scientist*, 28 July 1990; and Bill Sweetman, op cit, pp91–94.
39. The two coincidental events were later described in the *New York Times*, 5 June 1977.
40. According to Professor Gordon F. McDonald, associate director of the Institute of Geophysics and Planetary Physics at the University of California, Los Angeles, and member of the US President's Science Advisory Committee 1966.
41. *Science News*, 18 June 1994.
42. Nick Begich and James Roderick, *Earth Rising – The Revolution: Toward a Thousand Years of Peace*, Earthpulse Press, Anchorage, AK, 2000.
43. Figures from Gary T. Whiteford, 'Earthquakes and Nuclear Testing: Dangerous Patterns and Trends', presented at the Second International Conference on the United Nations and World Peace, Seattle, Washington, 14 April 1989.
44. 'Earthquakes Induced by Underground Nuclear Explosions: Environmental and Ecological Problems', edited by Rodolfo Console and Alexi Nikolaev, Springer-Verlag, Berlin, 1995 (published in cooperation with NATO).
45. *New York Times*, 1 March 1987.
46. P.A.C.E., *Newsletter*, vol 3, nos. 1–6, 1981.
47. P.A.C.E., *Newsletter*, vol 3, nos. 1–6, 1981. Located by Steve Elswick, Exotic Research, PO Box 5382, Security Colorado 80931, USA.
48. P.A.C.E., *Newsletter.*
49. *Newsweek*, 6 July 1993.
50. *Wall Street Journal*, 2 October 1992.
51. *Defense Daily Reports*, US air force, September and October 1994.
52. Alain-Claude-Galthe, 'Is El Niño Now a Man-Made Phenomenon?', *The Ecologist*, vol. 29, p64, 1999.
53. Interview in *Maclean* magazine, 5 August 1996.
54. *Toronto Star*, 9 July 1996.
55. C. Flavin, 'Facing Up to the Risks of Climate Change', Chapter 2 in *State of the World 1996, A Worldwatch Institute Report on Progress Toward a Sustainable Society*. Lester Brown et al. (eds), W.W. Norton & Co., New York, NY, 1996.
56. Jose Lutzenberger (president of Fundacio Gaia in Brazil), 'Gaia's Fever', *The Ecologist*, vol. 29, no. 2, 1999.

57. Mark Jaffe, 'What hath Man Wrought', The Frankline Institute Online.
58. Simon Retallack and Peter Bunyard, 'We're Changing our Climate! Who Can Doubt It?' *The Ecologist*, vol. 29, no. 2, 1999, p60.
59. 1946–63 for the Pacific and North America.
60. H.A. Bethe et al., 'Space-based Ballistic Missile Defense', *Scientific American*, vol. 2511, no. 4, 1984, p.37.
61. D.J. Kassler, 'Orbital debris issues', a paper presented at the COSPAR Congress, Graz, Austria, 1984.
62. See Frank Wentz and Matthias Schnabel, 'Effects of Orbital Decay on Satellite-Derived Lower-Tropospheric Temperature Trends', *Nature*, vol. 394, no. 6694 pp661–4, 1998.
63. *State of the World*, op cit, p27.
64. Steven Hume, in *The Vancouver Sun*, 30 December 1998.
65. *The Ecologist*, special issue on Climate Change, vol. 29, March/April 1999.

Chapter 5

1. *Humanitarian Times*, 5 April 2000.
2. *Guardian*, 1 December 1995, available also on line: *http://onli.guardian.co.uk/science/951201scexztiyah.html.*
3. Memo from the Catholic Fund for Overseas Development, the Catholic Institute for International Relations, Christian Aid, Oxfam, Save the Children Fund and the World Development Movement, sent to the Foreign Affairs Select Committee.
4. *The World Guide: 1999–2000*, Millennium Edition, The New Internationalist, Oxford, 1999.
5. US Office of Budgets projection for 2001.
6. J. Pike, 'US and Soviet Ballistic Missile Defence Programmes', in *Outer Space: A Source of Conflict or Cooperation*, Bhupendra Jasani (ed.), United Nations University Press, Tokyo, 1991.
7. Ruth Sevard, *World Military and Social Expenditure*, Washington, DC.
8. Ian Davis, 'Europe, Diversification or Conversion: More than Just Semantics?' Project on Demilitarisation, Leeds, England, in *Press for Conversion!* No. 24, February 1996.
9. Stockholm International Peace Research Institute Yearbooks 1992 to 1994.
10. US Bureau of Labour Statistics figures, publicised by The Campaign Against the Arms Trade, 5 Caledonian Road, London N1 9DX, UK.
11. This methodology assumes a global population of 5.892 billion, 1.8

hectares per capita available, and 2.3 hectares per capita actually being consumed.

12. The definition of biodiversity contained in the text of the Biodiversity Convention signed at the UN Conference on Environment and Development in Rio in 1992 is derived from growing concern over the rapid disappearance of species. It concentrates on resources reproduced in the wild, land and sea ecosystems, the potential use as medicines and their genetic resilience. Many ecologists believe that a larger percentage of the world's ecosystem needs to be preserved to secure biodiversity. In 1970, ecologist Eugum Odum recommended 40 per cent, and in 1991, Reed Noss, scientific director of the Wildlands Project, hypothesised that 50 per cent, on average, in a region needs to be protected as wilderness in order to restore large carnivores and meet other well-recognised conservation goals.

13. *A Guide to the Global Environment 1996–1997*, a joint publication of the World Resources Institute, the United Nations Environment Program, the United Nations Development Program and the World Bank, Oxford University Press, New York, 1996.

14. NATO Committee on the Challenges of Modern Society, cited in *The World Guide: 1999–2000*, Millennium Edition, The New Internationalist, Oxford, 1999.

15. R. Costanza, et al., 'The Value of the World's Ecosystem Services and Natural Capital', *Nature*, vol. 387, no. 6630, pp235–260, 15 May 1997.

16. Copies of these documents can be found in the United Nations archives. Because I was involved in the process, I also received copies of the papers with rings round the contested points.

17. Years earlier I had identified the issues that were debated at this summit in my book *No Immediate Danger: Prognosis for a Radioactive Earth*, The Women's Press, London, 1985.

18. M. Wackernagel, et al., 'Ecological Footprint of Nations', Centro de Estudios para la Sustentabilidad, Universidad Anahuac de Xalapa, Apartado Postal 653, 91000 Xalapa, Ver., Mexico, 10 March 1997.

19. Twenty main resources were analysed to determine consumption, defined to be production plus imports minus exports. These figures determined the national consumption of each resource. Using Food and Agriculture Organization (FAO) estimates of world average yield, consumption and waste absorption, these figures can be translated into hectares of land or sea required to meet consumption needs. Energy balance for each country was adjusted to include the energy

used for exported goods (consumed by other countries) and that required for imported finished products. The researchers calculated per capita consumption figures so as to provide a fair comparison between large and small countries. The yield factors probably overestimate the biological productivity of industrialised countries with heavy fertiliser use. Due to the international nature of the oceans, sea space was allocated equally to all nation's citizens.

20. See Paul Hawken, Amoury Lovins and L. Hunter Lovins, *Natural Capitalism – The Next Industrial Revolution,* Earthscan, London, 1999.

21. M. Wackernagel et al., op cit.

22. *A Guide to the Global Environment,* 1996–97, op cit.

23. Carbon Dioxide Information Analysis (CDIAC), Oak Ridge National Laboratory, '1992 Estimates of CO_2 Emissions from Fossil Fuel Burning and Cement Manufacturing Based on United Nations Energy Statistics and the US Bureau of Mines Manufacturing Data', ORNL/CDIAC-25, NDP-030 (an accessible database) Oak Ridge, Tennessee, September 1995.

24. Ernst Ulrich von Weizacker, *Factor Four,* Amory Lovins and L. Hunter Lovins, Earthscan, London, 1997.

25. 'Veterans and Agent Orange Update 1996', Committee to Review the Health Effects in Vietnam Veterans of Exposure to Herbicides, Division of Health Promotion and Disease Prevention, Institute of Medicine, US National Academy Press, Washington DC.

26. '1.2 Million Award in Agent Orange Suit', AP, *New York Times,* 27 May 1996.

27. Rajiv Chandrasekaran, 'War's Toxic Legacy Lingers in Vietnam', *Washington Post* Foreign Service, 18 April 2000.

28. The Superfund began with the passage of the Comprehensive Environmental Response, Compensation and Liability Act (CERCLA) by the US Congress in 1980. It was modified in 1986 by the Superfund Amendments and Re-authorization Act (SARA) which committed over $15 billion to clean up more than 2000 polluted sites recognised as posing a severe threat to human health. The average clean-up of one site has been estimated at $30 million.

29. Rare earth elements are a series of chemical elements with atomic numbers 57 to 71. They generally occur together and are difficult to separate from each other.

30. See Civil Suit No. 185 of 1985, in the High Court of Malaya at Ipoh, between Woon Tan Kan (deceased) et al. and Asian Rare Earth

Sendirian Berhad. Court Order 14 October 1985 to stop operations. The Court papers in support of this injunction include the recommendations of the I.A.E.A.

31. I attended almost all of the court sessions over this two-year period and coached the lawyers working for the people.

32. Among the trade names for chlordimeform are Acaron, Bermat, Bermachlorfenamidine, C 8514, Chlordimeform, Chlorophenamide, Ciba 8514, Ciba-C 8514, Cotip 500 EC, ENT 27567, EP-333, Fundal 500, Fundex, Galecron, ovinaovitix, Ovitoxionschhering-36268, RS 141, Schering 36268, Spanon and Spanone.

33. *Return to the Good Earth; A Third World Network Dossier*, Third World Network, Penang, Malaysia. See also 'Dangerous Exposures' *India Today*, 15 January 1985.

34. T.R. Chouhan et al., *Bhopal: The Inside Story*, with an afterword, 'Bhopal Ten Years After', by Claude Alvares and Indira Jaising, published in cooperation with the International Coalition for Justice for Bhopal by The Apex Press/The Other India Press, Mapusa 403 507 Goa, India, 1994.

35. Peter Snell and Kirsty Nicol, 'Pesticide Residues in Food: The Need for Real Control', Report of the London Food Commission quoted in '49 Pesticides in Link with Cancer, Report Claims', James Erlichman, *Guardian,* 4 March 1986.

36. Amitya Baviskar and Chiranjeev Bedi, Third World Network, 'Why Pesticides Can Never be Safe' (see ref 33 above).

37. Beth Hanson, 'Spoiled Soil', *Amicus* Journal, Summer 1989.

38. Rachel Carson, *Silent Spring*, Houghton Mifflin, New York, NY, 1962.

39. Robert Rudd, *Pesticides and the Living Landscape*, quoted in 'The Witch-hunt of Rachel Carson', by Frank Graham Jr., *The Ecologist*, vol. 10, no. 3, March 1980.

Chapter 6

1. Joanne Macy, *Despair and Personal Power in the Nuclear Age*, New Society Publishers, Philadelphia, PA, 1983.

2. Information on local, national and international peace organisations can be obtained from: International Peace Bureau, 41 rue de Zurich, CH-1201 Geneva, Switzerland; Tel: +41-22-731-6429; Fax: +41-22-738-9419.

3. 'Peril Seen in Globalizing Arms Industry', *Disarmament Times*,

published by the NGO Committee on Disarmament, 777 UN Plaza, New York, NY 10017, September 1999, p1.

4. Pugwash Conference on Science and World Affairs is a good peace contact for scientists and engineers. They have three offices, in Geneva, London and Rome, and many national chapters. The London office can be contacted at: Flat A Museum Mansions, 63A Great Russell Street, London WC1B 3BJ, England, UK; tel: +44-20-7405-6661; fax: +44-20-7831-5651; email: *pugwash@qmw.ac.uk.*

5. See Richard Deats, 'The Global Spread of Active Nonviolence', *Fellowship of Reconciliation,* vol. 62, July/August 1996; and Arthur Laffin and Anne Montgomery (eds) *Swords Into Plowshares: Nonviolent Direct Action for Disarmament, Peace and Social Justice,* revised edition, Fortkamp Publ., Rose Hill Books, 28291-444th Avenue, Marion, SC 57043, 1996. This is a collection of essays from many of the leading peacemakers of our day, together with comments covering wars as recent as that in Iraq.

6. Don Kraus, 'Most Americans deplore US tactics', *Mondial,* Journal of the World Federalists Canada, January 2000.

7. Fergal Keane, BBC African reporter, quoted in Nicholas Hildyard and Sarah Sexton, '"Blood", "Culture" and Ethnic Conflict', Cornerhouse Briefing Papers, The Corner House, Sturminster, Newton, UK, January 1999. *cornerhouse@gn.apc.org*

8. 'Time for Truly Ethical Policies', *Guardian,* 12 June 2000.

9. 1992 Pentagon Defense Planning Guide.

10. This report is available on the Internet at *http://www.dfat.gov.au/dfat/cc/cchome.html,* or it can be obtained from the NGO Committee on Disarmament, 777 United Nations Plaza, New York, NY 10017.

11. *NATO Review,* July–August 1997, p9.

12. Ibid, p8.

13. 'NATO to Review Nuclear Weapons Policy', *Mondial,* Journal of the World Federalists Canada, January 2000.

14. Figures from 'Hidden Killers: The Global Demining Crisis', US Department of State, Washington DC, publication no. 190575, 1998. The numbers represent rounded estimates.

15. Quoted in Silvija Jaksic, 'Landmines', *Peace Magazine,* official magazine of Science for Peace Canada, July/August 1996.

16. The US refusal to sign the landmine ban was opposed by the Vietnam Veterans of America Foundation, RR 1, Box 871, Putney, Vermont 05346; tel: +1-802-387-2080; fax +1-802-387-2081; email: *banmines@sover.net* and by other US non-governmental organisations.

17. UN General Assembly Resolution, 2 November 1994.
18. International Court Document No. 96/23, 8 July 1996.
19. This third decision was by a split vote. In favour: President Bedjaoui, Judges Ranjeva, Herczegh, Shi, Fleischhauer, Vereschetin, Ferrari Bravo; Against: Vice-President Schebel and Judges Oda, Guillaume, Shahabuddeen, Weeramantry, Koroma, Higgins.
20. The International Court of Justice, Peace Palace, The Hague, Communique No. 96/23, 8 July 1996.
21. Normally legal statements are not allowed to be compound for the reason that a 'yes' or 'no' vote on a compound statement can lead to more than one interpretation.
22. Ernie Regehr, President of Project Ploughshares, 'Canadian Non-Proliferation Treaty Delegation Report to Non-Governmental Organizations #4', 16 May 2000.
23. A coalition of NGO organisations under the title Abolition 2000, with contact groups in many countries, has been formed to keep up the international pressure for nuclear disarmament. An appendix of peace groups appears at the end of this book.
24. Statement made at the 12 July 1996 Brussels Meeting of the International Peace Bureau.
25. One international NGO involved in this area is Global Network Against Weapons & Nuclear Power in Space, PO Box 90083, Gainesville, FL 32607; email: *globalnet@mindspring.com*; web page: *www.globenet.free-online.co.uk.*
26. Betty Reardon, 'Gender and Global Security: A Feminist Challenge to the United Nations and Peace Research', *Journal of International Cooperation Studies*, vol. 6, June 1998.
27. International Fellowship of Reconciliation, Spoorstraat 38, 1815 BK Alkmaar, The Netherlands.

Chapter 7

1. At the time Boutros Boutros-Ghali was secretary general of the United Nations.
2. Unicef Report, 30 April 1998.
3. These countries were Iraq, Yugoslavia, Libya, Liberia, Somalia, Haiti, Angola, Rwanda, Sudan, Cambodia, Sierra Leone, and Afghanistan.
4. J. Tuyet Nguyen, '"Smart sanction" needed, Axworthy Tells UN Body', *Toronto Star*, 18 April 2000, A16.
5. S. Schmidheiny et al., 'Signals for Change: Business Progress Towards

Sustainable Development', World Business Council for Sustainable Development, E&Y Direct, PO Box 934, Poole, Dorset, BH17 7BR, tel: +44-1202-679-885; fax: +44-1202-661-999.

6. With a view to explain in easy terms what is at stake for the environment and sustainable development in the next round of WTO negotiations, World Wildlife Fund has prepared an information/ lobbying pack for NGOs, parliamentarians, journalists, government officials and other people interested in the issue (initially in English, French and Spanish). As Claude Martin, Director of WWF International, writes in an introductory letter to the pack, 'This information pack is designed to explain the link between trade, the environment and the need to build a sustainable world economy. It also enables those of you who wish to play a part in reshaping international trade to protect all our futures.' Mikel Insausti, WWF EPO, 'Sustainable Trade for a Living Planet', can be requested from Flo Danthine, WWF European Policy Office, email: *<Fdanthine@wwfnet.org>* mail to: *Fdanthine@wwfnet.org*; or fax: +32.2 743.88.19.

7. 'War, Lies and Videotape: How the Media Monopoly Stifles the Truth', based on international conference held in Greece, 24–28 May 1998, by Women for Mutual Security, published by International Action Center, New York, NY, 2000.

8. The Venice Declaration of the IV International Conference, 'Toward the World Governing of the Environment', Cini Foundation, Venice 2–5 June 1994. Can be obtained from International Court of the Environment Foundation (ICEF), Palazzo di Guistizia, Piazza Cavour, 1, 00193 Roma, Italy.

9. Women's Planet, Via S. Maria dell'Anima, 30, 00186 Rome, Italy.

10. WEDO, 355 Lexington Avenue, Third Floor, New York, NY 10017-6603, USA.

11. The Secretariat for the International Court of the Environment Foundation (ICEF) is in the Palazzo di Guistizia, Piazza Cavour, 1, 00193 Rome, Italy.

12. 'Senate Committee Reports Detail Requirement for UN Reform' Washington Report, 6 May 1999. Available from the United Nations Association (UNA) of USA, Washington Office.

13. The Earth Council is also supported by private donors.

14. International Secretariat, Earth Council, San Jose, Costa Rica; tel: +506-256-1611; fax: +506-255-2197; email: *info@terra,ecouncil.ac.cr*; Web: *http://www.ecouncil.ac.cr.*

15. Some of the problems with global development are highlighted in the journal *Development Alternatives*, B-32 Tara Crescent, Qutab Institutional Area, New Delhi 110 016, India.

16. Andre de Moor and Peter Calami, 'Subsidizing Unsustainable Development: Undermining the Earth with Public Funds', published by the Earth Council, San Jose, Costa Rica, 1997.

17. Originally entitled *Pocket Manual of Rules for Deliberative Assemblies* (1876), *Robert's Rules of Order* is regarded in the United States as the authoritative statement of parliamentary rules governing public meetings. The first edition was written by Henry Martyn Robert, an army engineer, and was essentially an adaptation of the practice of the US House of Representatives.

18. The International Environmental Agency for Local Governments, also known as the International Council for Local Environmental Iniatives (ICLEI), World Secretariat, City Hall, 8th floor, East Tower, Toronto, Ontario, Canada M5H 2N2. The European Secretariat is located at Eschholzstradde 86, D-79115 Freiburg, Germany; the Asian Pacific Secretariat is located at c/o GEF, Iikura Building, 1-9-7 Azabudai, Minato-ku, Tokyo 106 Japan; the Office of the African Regional Coordinator is at PSA House, 9 Livingstone Avenue, PO Box 6852, Harare, Zimbabwe; the Office of the Latin American Regional Coordinator is at Corporacion para el Desarrollo de Santiago, Av. Cardenal Jose Maria Caro 390, PO Box 51640, Correo Central, Santiago, Chile; the US Cities for Climate Protection is at 15 Shattuck Square, Suite 215, Berkeley, CA 94704, USA.

19. L.J. Onisto, E. Krause and M. Wackernagel, 'How Big is Toronto's Ecological Footprint', the Centre for Sustainable Studies and the City of Toronto. September 1998.

20. Hans-Peter Durr, Max Plank Institute fur Physik, Fohringer Ring 6, D-80805 Munich, Germany.

21. WHO Report of toxic waste left on military bases in the Philippines by US Armed Forces, May 9, 1993.

22. Dr. Rosalie Bertell, 'Health for All: A Study of the Health of People Living on or near to the Former US Clark Air Force Base 1996–1998', a joint project of the International Institute of Concern for Public Health, Toronto, and People's Task Force for Base Cleanup, Manila. Published by the IICPH, Toronto, Canada, October 1998.

23. 'Military Base Closures, US Financial Obligations in the Philippines', US General Accounting Office, 1992.

24. M. Minkler (ed), *Community Organizing and Community Building for*

Health, Rutgers University Press, New Brunswick, NJ, and London, 1997. See also *Statistics Needed for Determining the Effects of the Environment on Health*, Report of the Technical Consultant Panel of the United States National Committee on Vital and Health Statistics, Series 4, Number 20. US Department of Health, Education and Welfare DHEW Publ. No. (HRA) 77-1457, 1977.

25. Great Lakes Research Consortium, State University of New York, College of Environmental Science and Forestry, Syracuse, New York.

26. These documents and 'Summary: State of Knowledge Report on Environmental Contaminants and Human Health in the Great Lakes Basin' are available from Environmental Health Effects Division, Wing 1100, Main Statistics Building, P.L. 0301 A 1, Tunney's Pasture, Ottawa, Ontario K1A 0K9, Canada.

27. *The Role of the Media in Health Risk Perception:A Literature Review*, prepared by Lori Abbott, for Great Lakes Health Effects Programme, Health Canada, August 1994.

28. INCHES reflects the perspectives of a wide spectrum of professions on the relationships between environment health and children. Contact person is Peter van den Hazel, MD, c/o Dutch Association of Environmental Medicine, PO Box 389, 6800 AJ Arnhem, The Netherlands.

29. Rosalie Bertell, 'Environmental Influences on the Health of Children', in *Risks, Health and Environment*, M.E. Butter (ed), Report No. 52, Science Shop for Biology, University of Groningen, The Netherlands, 1999.

30. *The State of the World's Children 1995*, James P. Grant (ed), executive director of Unicef, Oxford University Press, Oxford, 1996.

31. Rosalie Bertell, *No Immediate Danger: Prognosis for a Radioactive Earth*, The Women's Press, London, 1986.

32. Joby Warrick, 'US Plans to Pay for Ills from Radiation', *Washington Post*, 12 April 2000, A1.

33. The US National Academy of Sciences Publications on the Biological Effects of Ionizing Radiation, known as the BEIR Reports. Published by the US National Academy Press, Washington DC.

34. Joby Warrick, op cit.

35. Grant, op cit, Table 10, pp84–85.

36. 'UNEP and Habitat Welcome Group of Eight Cologne Debt Relief Initiative', UNEP Press Release, 21 June 1999.

37. Reported in *Asia Week*, 1 September 1995. This was before the Asian 'crash', but still indicates the influence of the wealthier investment

community in manipulating world markets and affecting significantly quality of life for ordinary people.

38. Based on an updated Report of Canada to the United Nations Commission of Sustainable Development, Third Session of the Commission, 11–28 April 1995. Available from the Department of Foreign Affairs and International Trade, Ottawa.

39. Alice Tepper Marlin, president of CEP, can be reached at 30 Irving Place, New York, NY 10003-2386, USA.

40. ISEE can be reached at School of Public Affairs, University of Maryland, College Park, MD 20742-1821, USA.

41. Committees of Soldiers' Mothers of Russia, 4 Luchnikov Lane, Door 3, Room 32, 103982 Moscow, Russia.

42. Prof. P.K. Ravindrian is president of KSSP. They can be reached at Kerala Sastra Sahitya Parishad, AKG Road, PO Edappally, Kochi 682 024, India.

43. SAM, 27 Lorong Maktab, 10250 Penang, Malaysia. The Third World Network and the Third World Health Network can be reached at: CAP-Malaysia, 228 Macalister Rd, Penang, 10250 Malaysia.

44. Manfred Max-Neef can be reached at Universidad Austral de Chile, Casilla 567, Valdivia, Chile.

45. Future in Our Hands, Torggata 35, 0183 Oslo 1, Norway.

INDEX

send for a free catalogue of all our titles

BLACK ROSE BOOKS

C.P. 1258, Succ. Place du Parc

Montréal, Québec

H3W 2R3 Canada

or visit our web site at: http://www.web.net/blackrosebooks

To order books in North America:

(phone) 1-800-565-9523 (fax) 1-800-221-9985

In the UK & Europe: (phone) 44(0)20 8986-4854 (fax) 44(0)20 8533-5821

Printed by the workers of

MARC VEILLEUX IMPRIMEUR INC.

Boucherville, Québec

for Black Rose Books Ltd.